Amazing

AREA MAZES

70 Race-the-Clock Puzzles for Budding Math Wizards

NAOKI INABA and RYOICHI MURAKAMI

NEW YORK

Originally published in Japan as *Area Mazes: Speed Edition* (面積迷路 スピード編) by Gakken Publishing Co., Ltd., Tokyo, in 2013. First published in North America in revised form by The Experiment, LLC, in 2019. English translation rights arranged with Gakken Plus Co., Ltd., through Paper Crane Agency.

The Experiment, LLC
220 East 23rd Street, Suite 600
New York, NY 10010-4658
theexperimentpublishing.com

ISBN 978-1-61519-618-0

Cover design by Sophie Appel
Series design by Sarah Schneider
Text design by Karen Giangreco
Translation by Erica Williams and Ayaka Takahashi | Paper Crane Editions

Manufactured in the United States of America

First printing September 2019
10 9 8 7 6 5 4 3 2 1

contents

INTRODUCTION

El Camino Training Academy for Science and Mathematics, which I direct, was established in 2006 in the hopes of creating an integrated curriculum for children from first to twelfth grades. We provide a unique science and mathematics curriculum that prepares students for school exams and for the Mathematical Olympiads. We use various puzzles in our classes for first to third graders, but area mazes (*menseki meiro*) are definitely among our best.

Our reasons for using puzzles in class are simple . . .

» Puzzles let children practice trial and error

Puzzles are a great way for today's kids to master the art of making mistakes, especially those kids who don't often make them, or are afraid to. With area mazes, the crucial first step is for children to get their pencils moving and simply try, try again. This kind of repeated trial and error is fundamental to performing science experiments—so area mazes can help students learn to think like scientists.

» Puzzles stretch children's abilities

The best way to improve (and speed up) calculation skills is simply through practice. Area mazes provide lots of arithmetic practice, and put these calculations in a meaningful context that children can see and understand (as opposed to leaving them in the abstract). Area mazes can also help develop children's spatial and logical reasoning, which are essential modes of scientific thinking.

» Puzzles encourage children to think for themselves

Children learn best when they are fascinated by a problem and eager to figure out a solution. Area mazes will entice children to invest time and concentration—while solving them creates a sense of achievement, giving children the chance to say, "I did it!"

That's why I ask this of parents, guardians, and teachers: Please don't give kids too many hints or tell them the solutions to the puzzles. Instead, ask kids to explain their thought processes to you each time they solve one. By explaining their thinking, children will take their comprehension to the next level.

When are kids ready for area mazes?

To solve area mazes, you need to know the area of a rectangle, which most children are taught in third grade. But these puzzles have been designed so that any child with basic knowledge of multiplication and division can solve them. Regardless of what grade your child might be in at school, if they show interest in area mazes then now is the best time to get them started.

It's true that kids are unlikely to encounter area mazes on their school exams. However, being exposed to high-quality math puzzles can help your child realize early on just how fun and interesting mathematics can be! I'm sure you'll agree once you see your child engaged and excited by these area mazes.

—**RYOICHI MURAKAMI**, director of El Camino

HOW TO SOLVE AREA MAZES

Using the given values, find (?). Remember, area = height × width.

If your calculation creates a fraction or decimal, STOP and look for another way. Area mazes are solved using whole numbers only! (However, do not assume that every value *in the diagram* must be whole.)

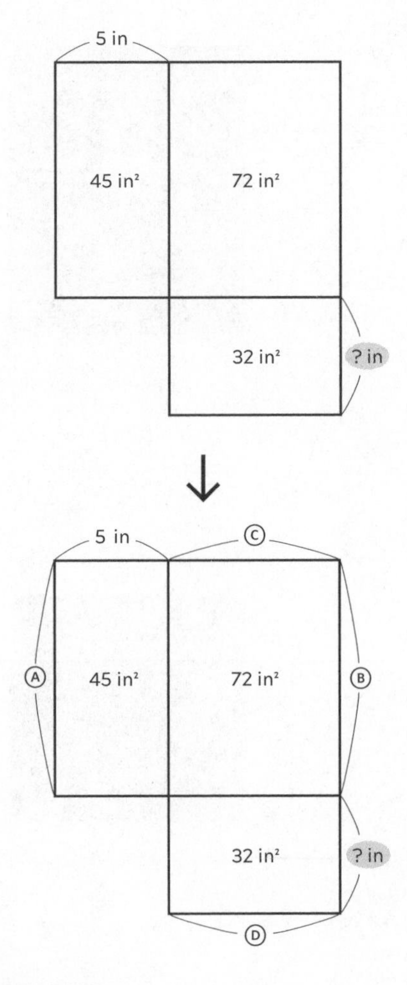

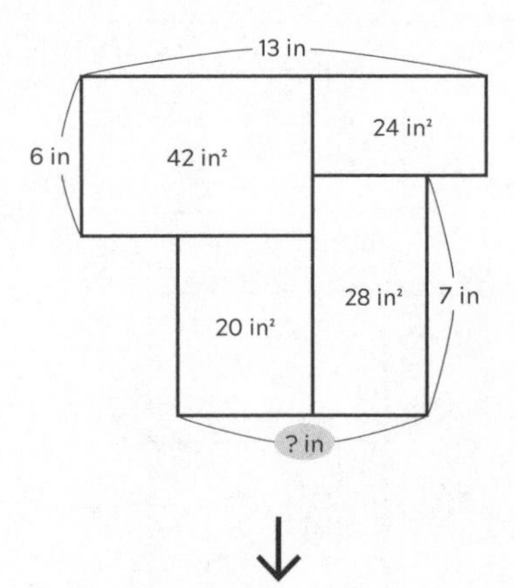

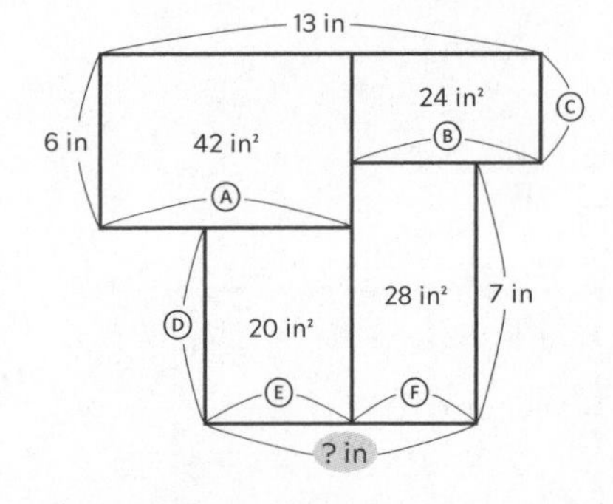

EXAMPLE ONE

Find Ⓐ . . . 45 ÷ 5 = 9 in.
Find Ⓑ . . . This is the same as Ⓐ, so 9 in.
Find Ⓒ . . . 72 ÷ 9 = 8 in.
Find Ⓓ . . . This is the same as Ⓒ, so 8 in.
Length (?) is 32 ÷ 8 = 4 in.

EXAMPLE TWO

Find Ⓐ . . . 42 ÷ 6 = 7 in.
Find Ⓑ . . . 13 – Ⓐ = 6 in.
Find Ⓒ . . . 24 ÷ Ⓑ = 4 in.
Find the total height of the puzzle . . .
Ⓒ + 7 = 11 in.
Use the total height to find Ⓓ . . .
11 – 6 = 5 in.
Find Ⓔ . . . 20 ÷ Ⓓ = 4 in.
Find Ⓕ . . . 28 ÷ 7 = 4 in.
Length (?) is Ⓔ + Ⓕ = 8 in.

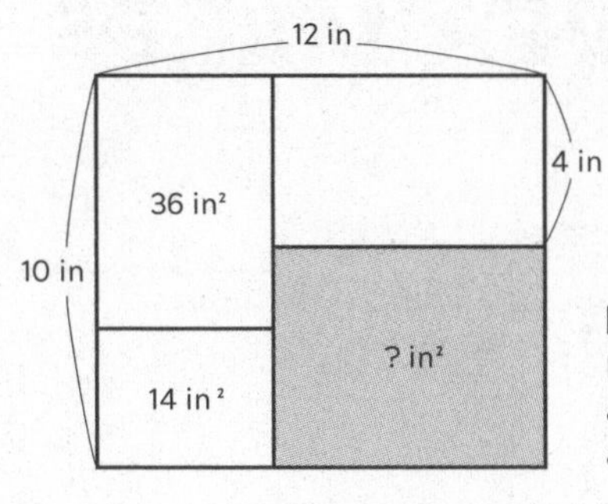

EXAMPLE THREE

First, create area ① by adding together the two areas on the left.

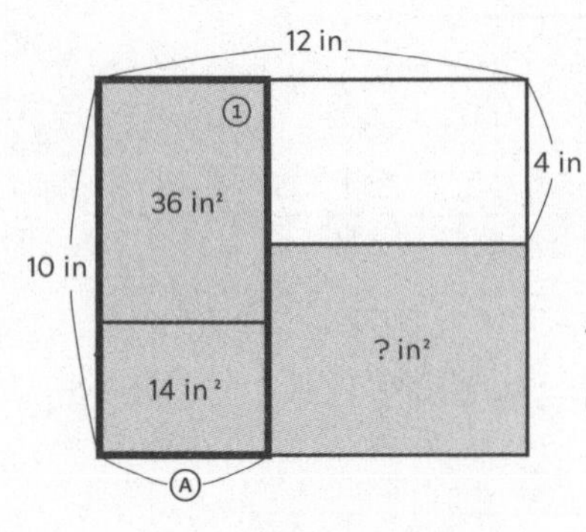

Area ① . . .
$36 + 14 = 50$ in.²
Use area ① to find Ⓐ . . .
$50 \div 10 = 5$ in.

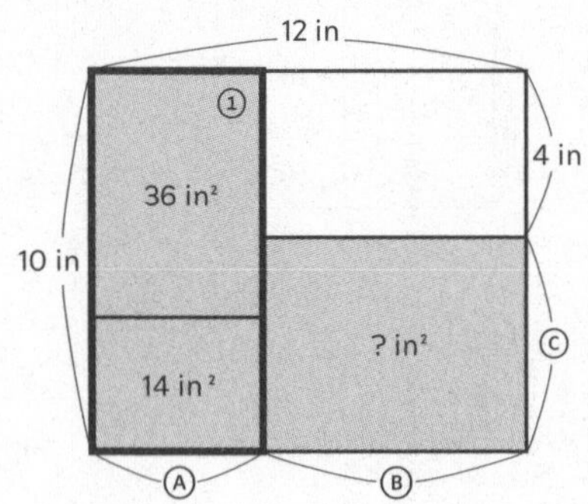

Ⓑ . . . 12 – Ⓐ = 7 in.
Ⓒ . . . 10 – 4 = 6 in.
Area ? is Ⓑ × Ⓒ = 42 in.²

Note that the figures are not drawn to scale. You can't solve by "eyeballing"—you have to prove it with math!

Even after you have solved a problem, you can revisit it to look for a more elegant solution.

warm-up puzzles

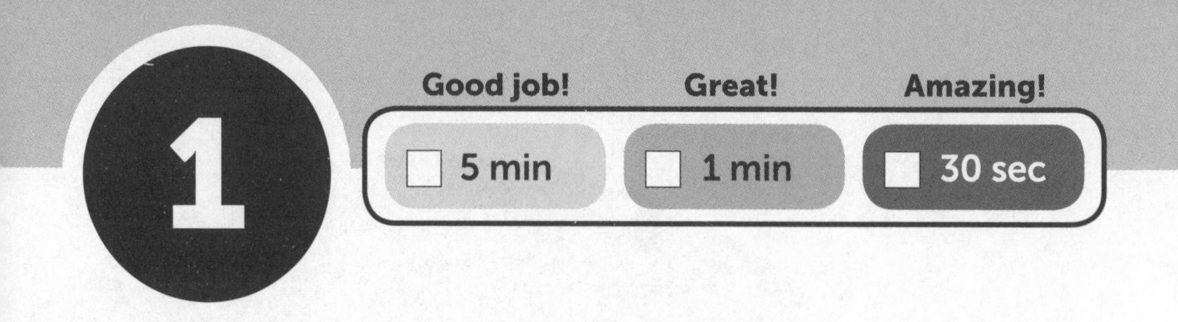

5 in

20 in²

16 in²

28 in²

42 in²

30 in²

? in

Solution

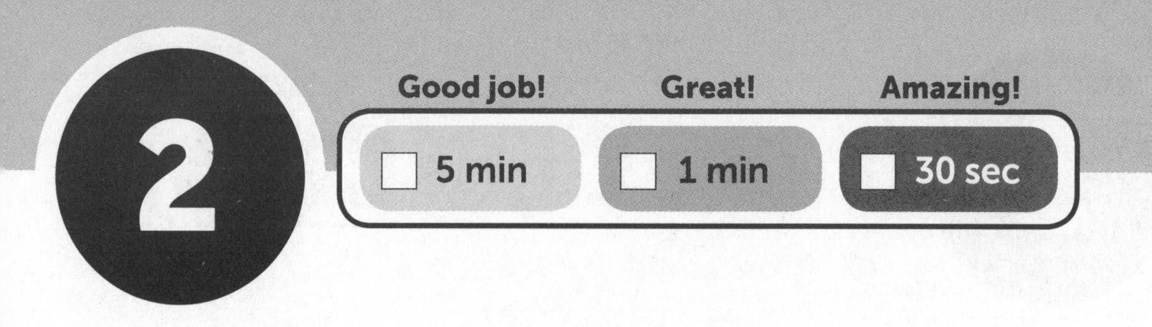

48 in²

40 in²

63 in²

? in

6 in

36 in²

25 in²

35 in²

Solution ____________

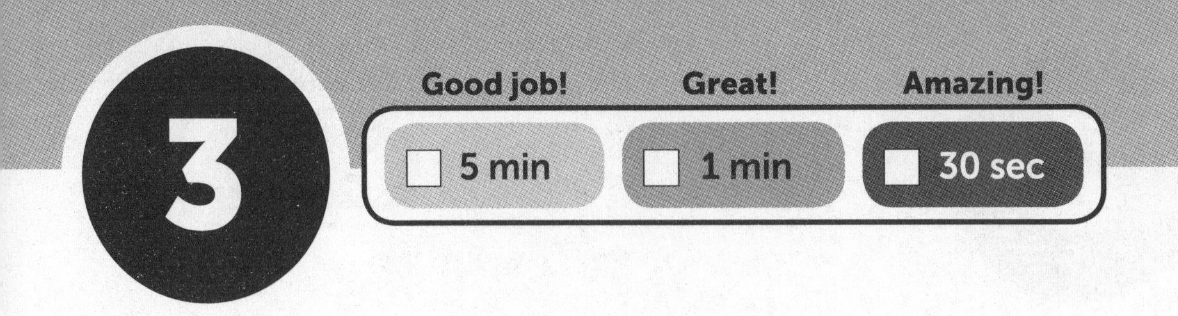

4 in	24 in^2	32 in^2
	54 in^2	$? \text{ in}^2$

Solution ______________

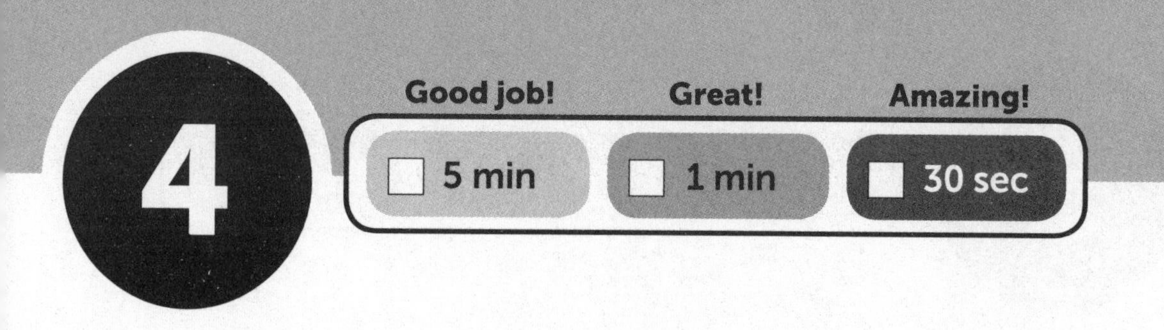

8 in

	56 in²	35 in²
48 in²		40 in²
54 in²	? in²	

Solution ______

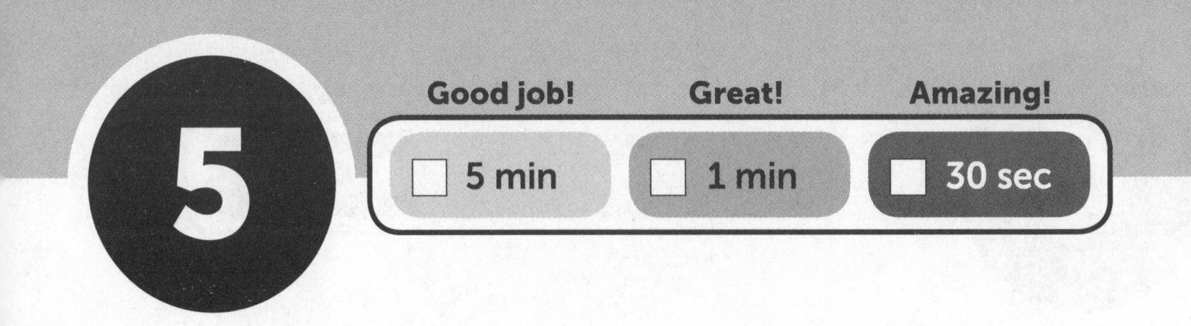
5
Good job!
Great!
Amazing!
5 min
1 min
30 sec

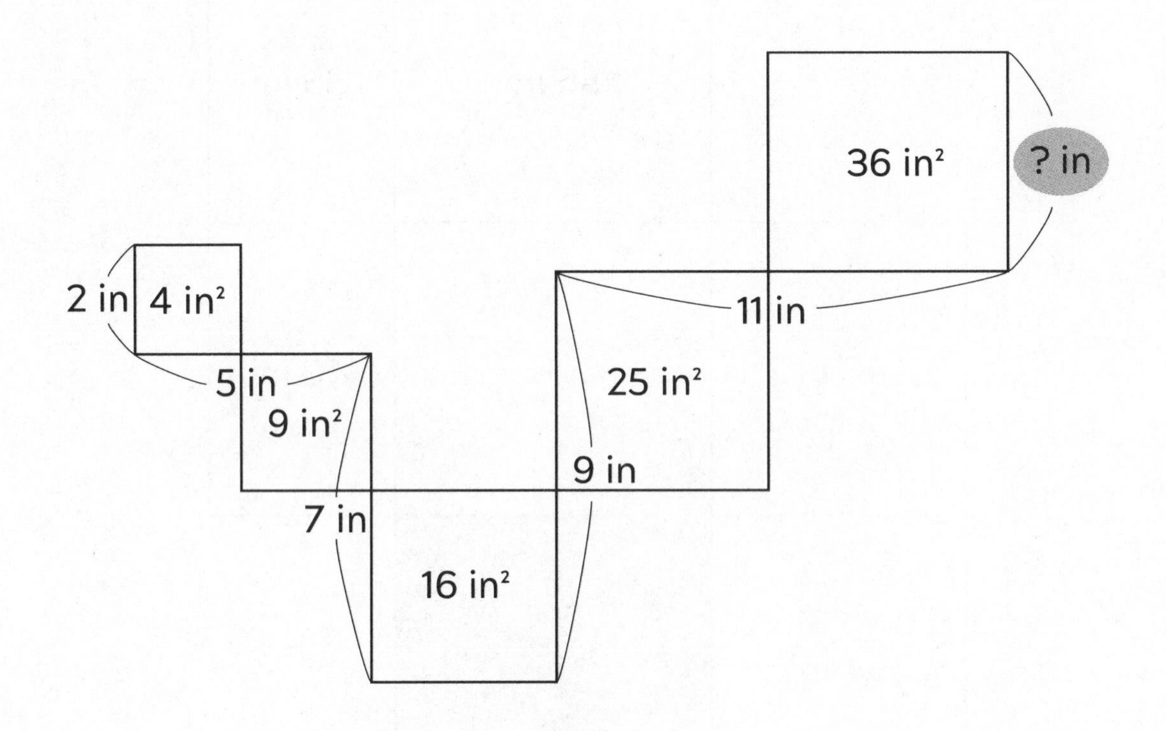
36 in²
? in
2 in
4 in²
11 in
5 in
25 in²
9 in²
9 in
7 in
16 in²

Solution

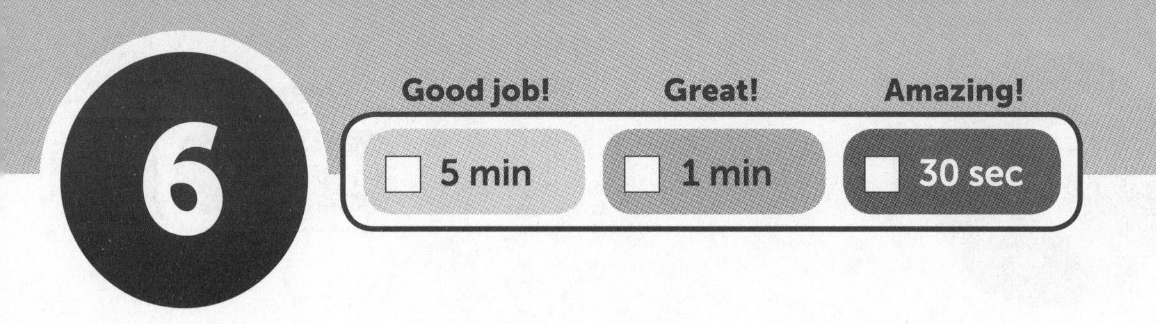
6
Good job!
Great!
Amazing!
5 min
1 min
30 sec

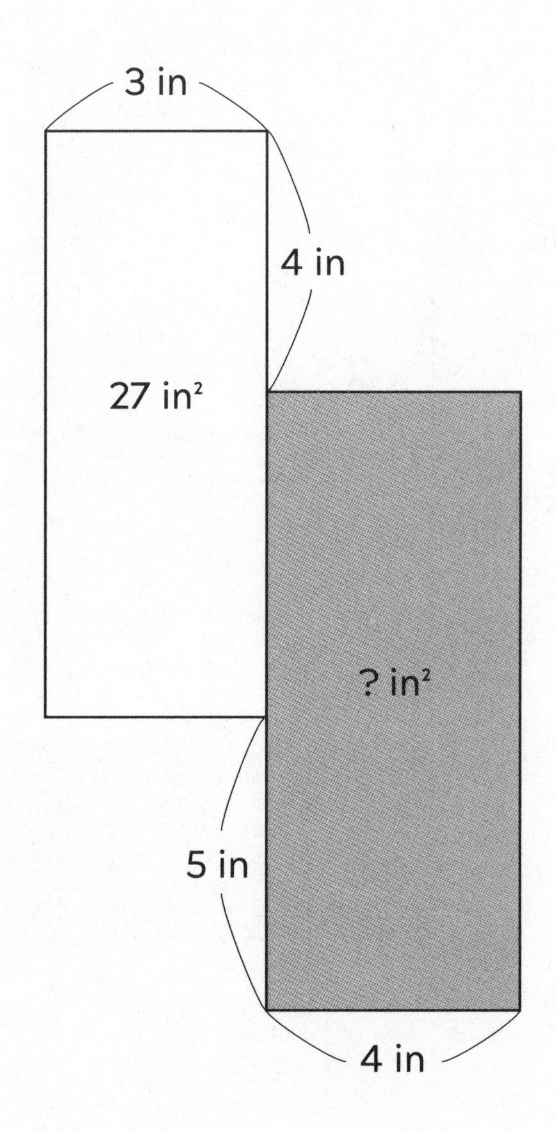
3 in
4 in
27 in²
? in²
5 in
4 in

Solution

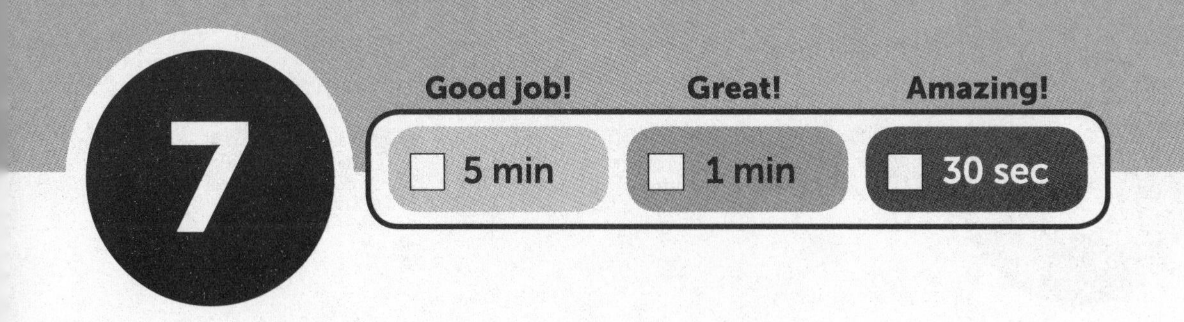
7
Good job!
Great!
Amazing!
5 min
1 min
30 sec

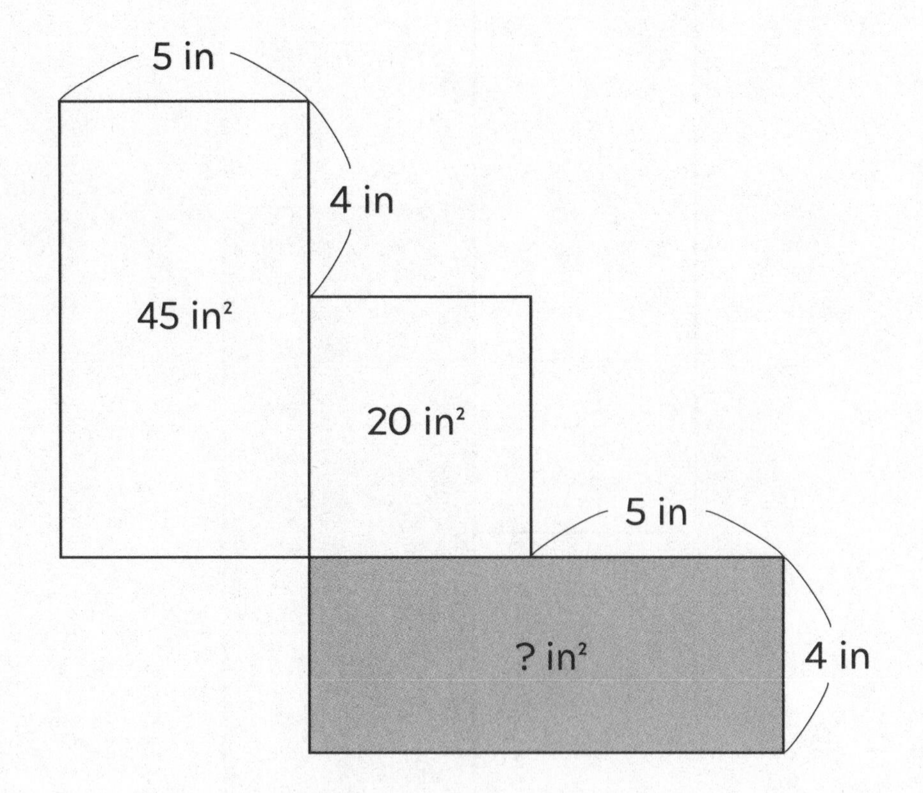
5 in
4 in
45 in²
20 in²
5 in
? in²
4 in

Solution ____________

skill-building puzzles

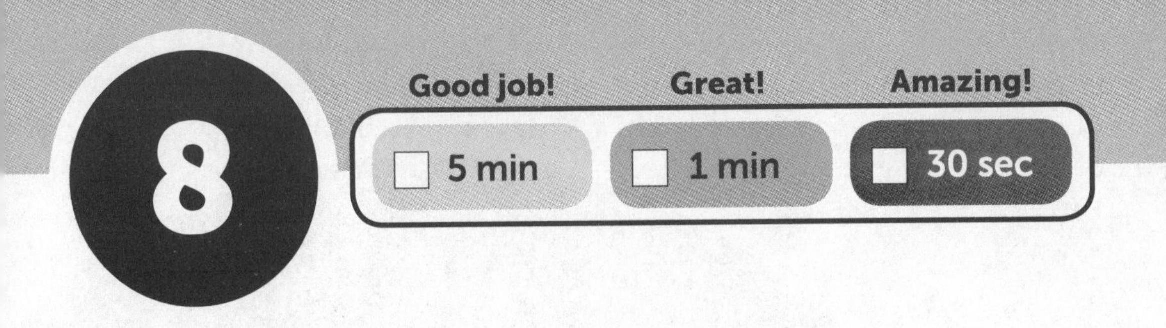
8
Good job!
Great!
Amazing!
5 min
1 min
30 sec

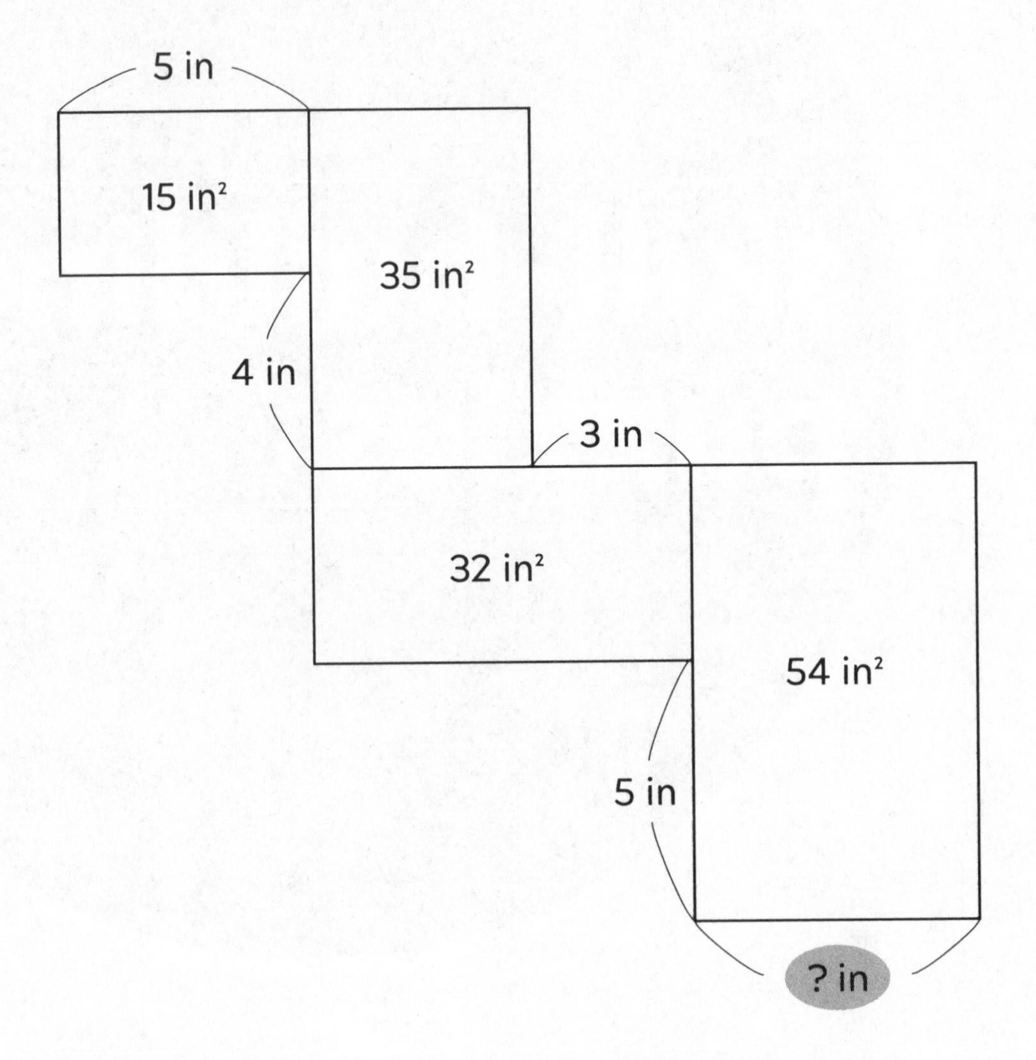
5 in
15 in²
35 in²
4 in
3 in
32 in²
54 in²
5 in
? in

Solution

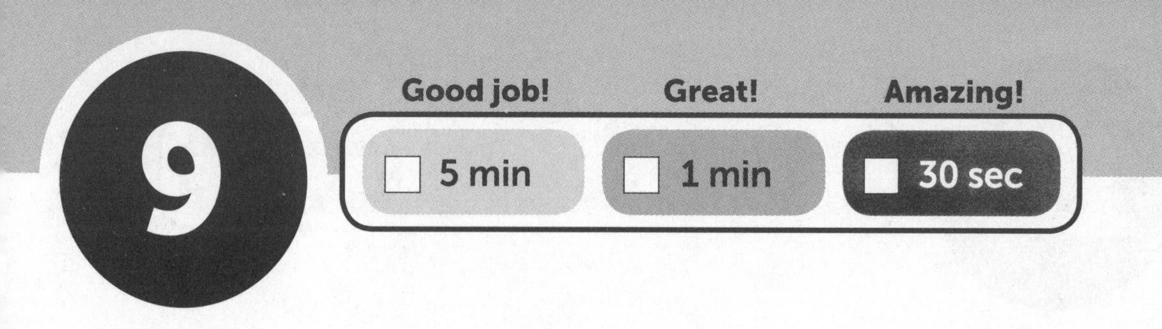
9
Good job!
Great!
Amazing!
5 min
1 min
30 sec

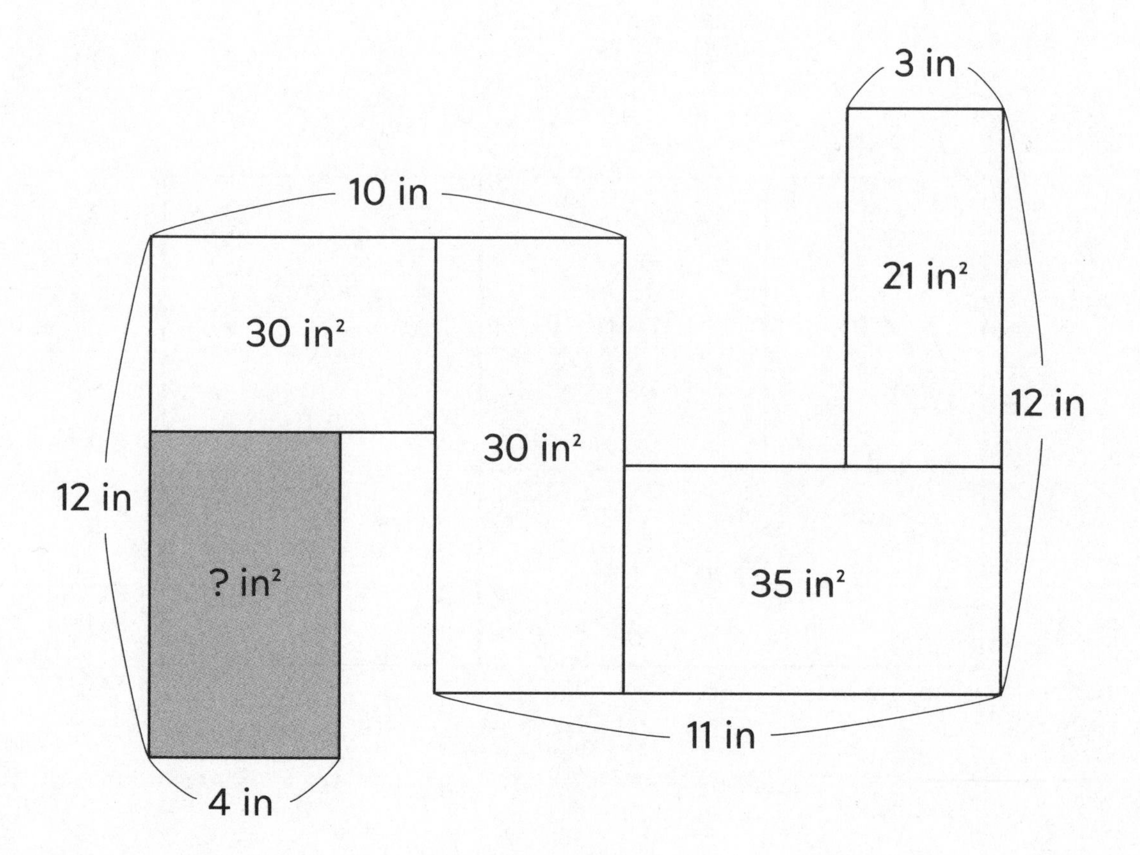
3 in
10 in
21 in²
30 in²
12 in
30 in²
12 in
? in²
35 in²
11 in
4 in

Solution ______

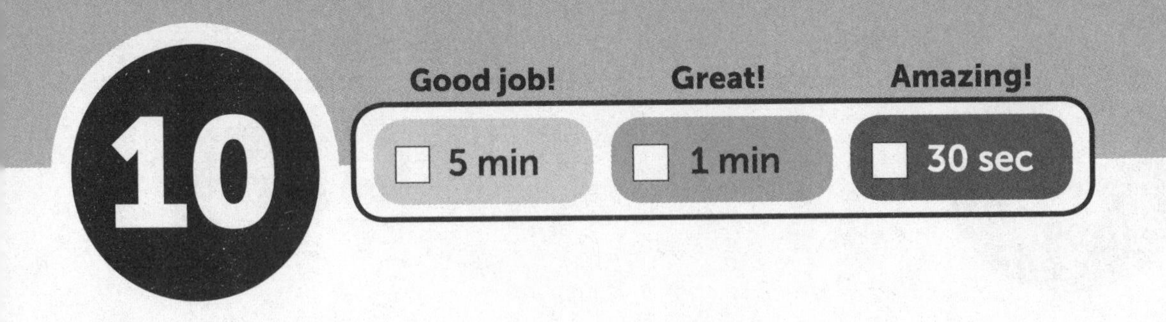
10
Good job!
Great!
Amazing!
5 min
1 min
30 sec

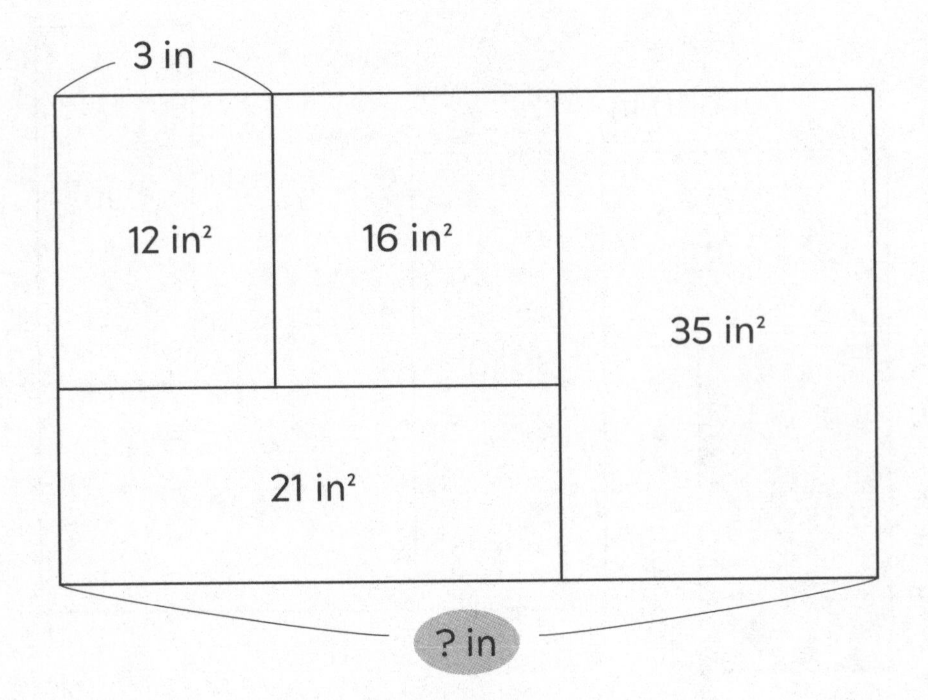
3 in
12 in²
16 in²
35 in²
21 in²
? in

Solution ____________

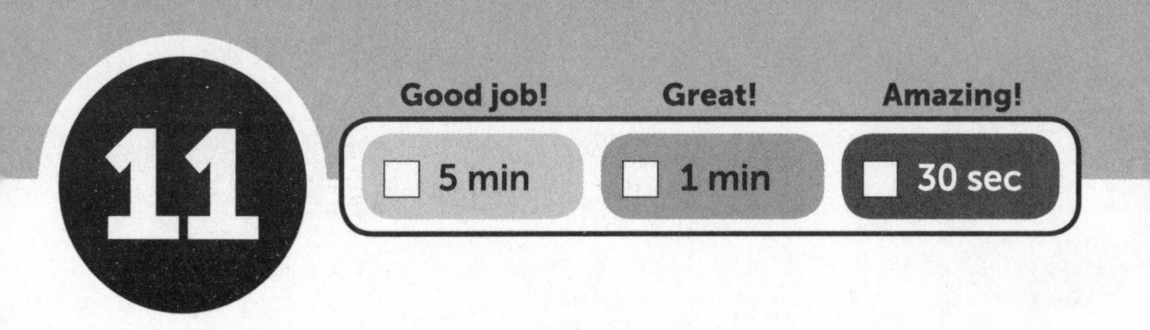
11
Good job!
Great!
Amazing!
5 min
1 min
30 sec

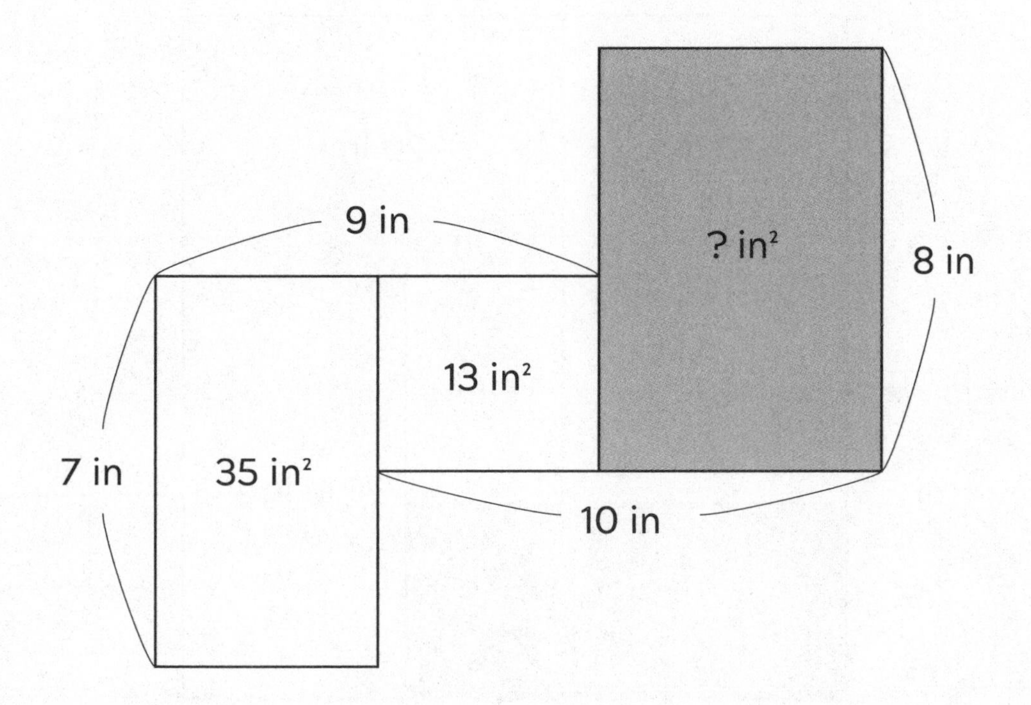
9 in
? in²
8 in
13 in²
7 in
35 in²
10 in

Solution

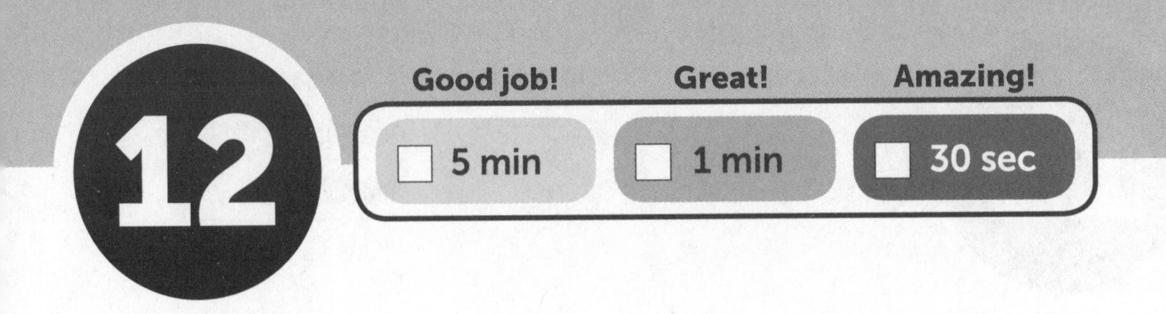

16 in^2

20 in^2

2 in

12 in^2

6 in^2

18 in^2

? in^2

Solution ______

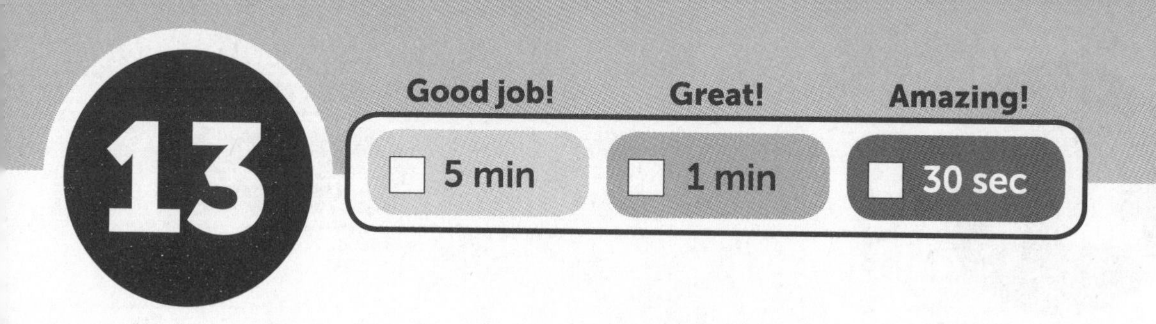
13
Good job!
Great!
Amazing!
5 min
1 min
30 sec

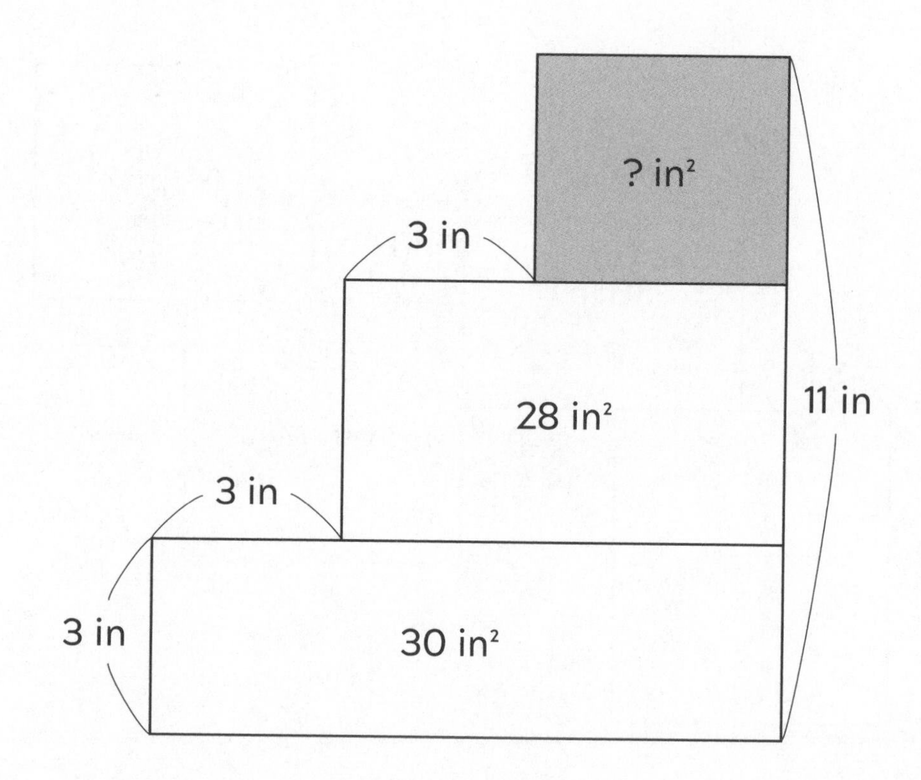
? in²
3 in
28 in²
11 in
3 in
3 in
30 in²

Solution ____________________

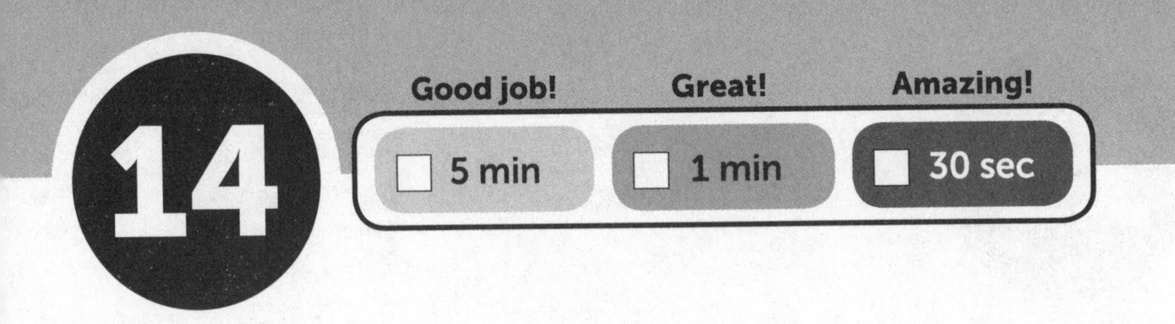
14
Good job!
Great!
Amazing!
5 min
1 min
30 sec

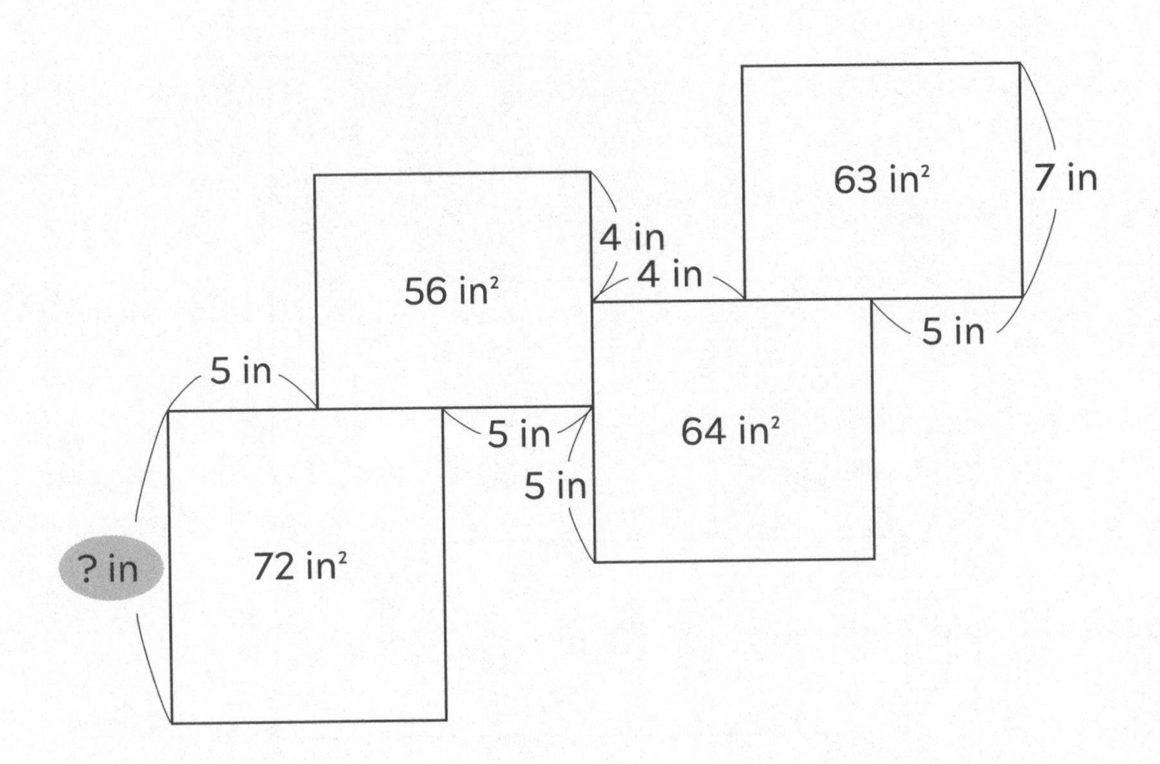
63 in²
7 in
56 in²
4 in
4 in
5 in
5 in
64 in²
5 in
5 in
? in
72 in²

Solution ______________

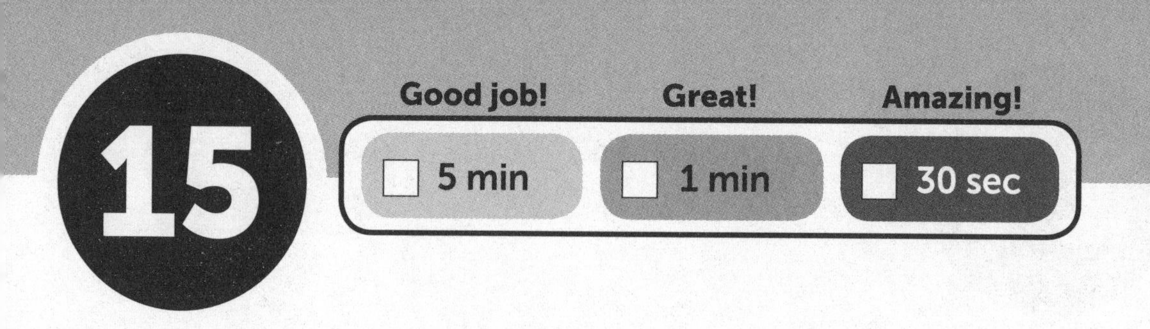
15
Good job!
Great!
Amazing!
5 min
1 min
30 sec

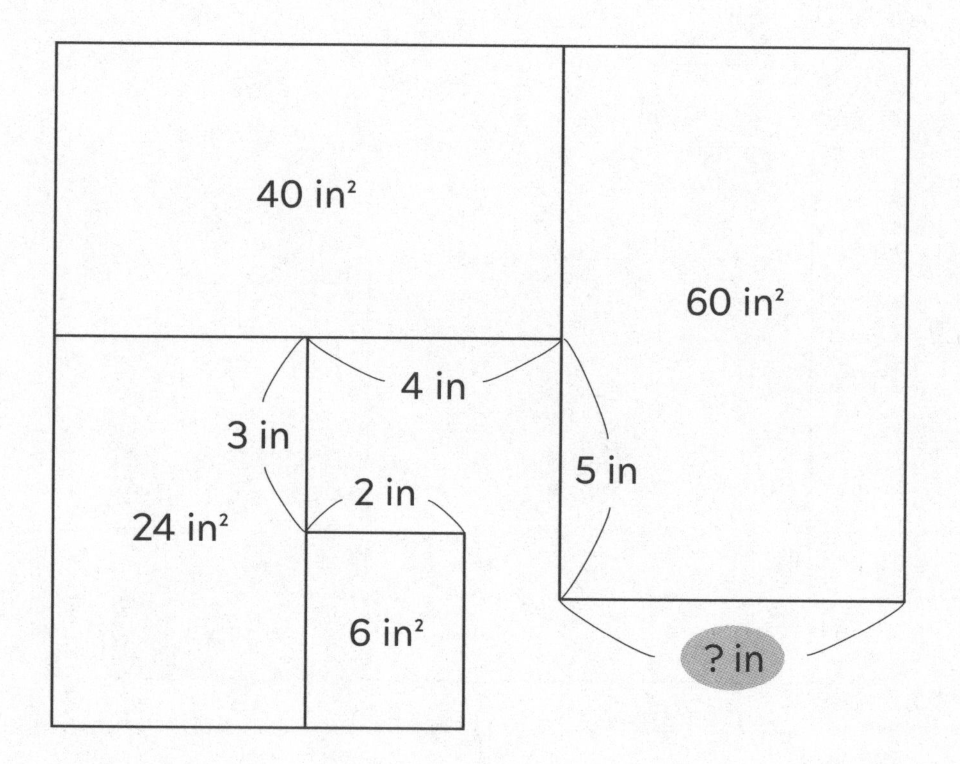
40 in²
60 in²
4 in
3 in
5 in
2 in
24 in²
6 in²
? in

Solution ______

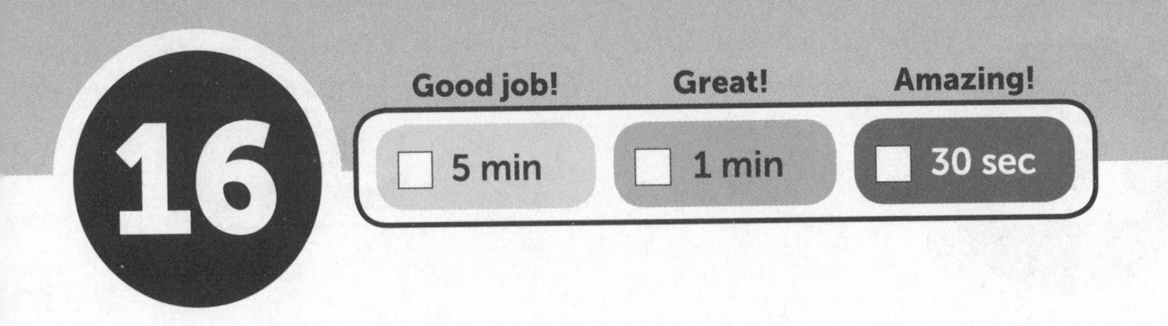
16
Good job!
Great!
Amazing!
5 min
1 min
30 sec

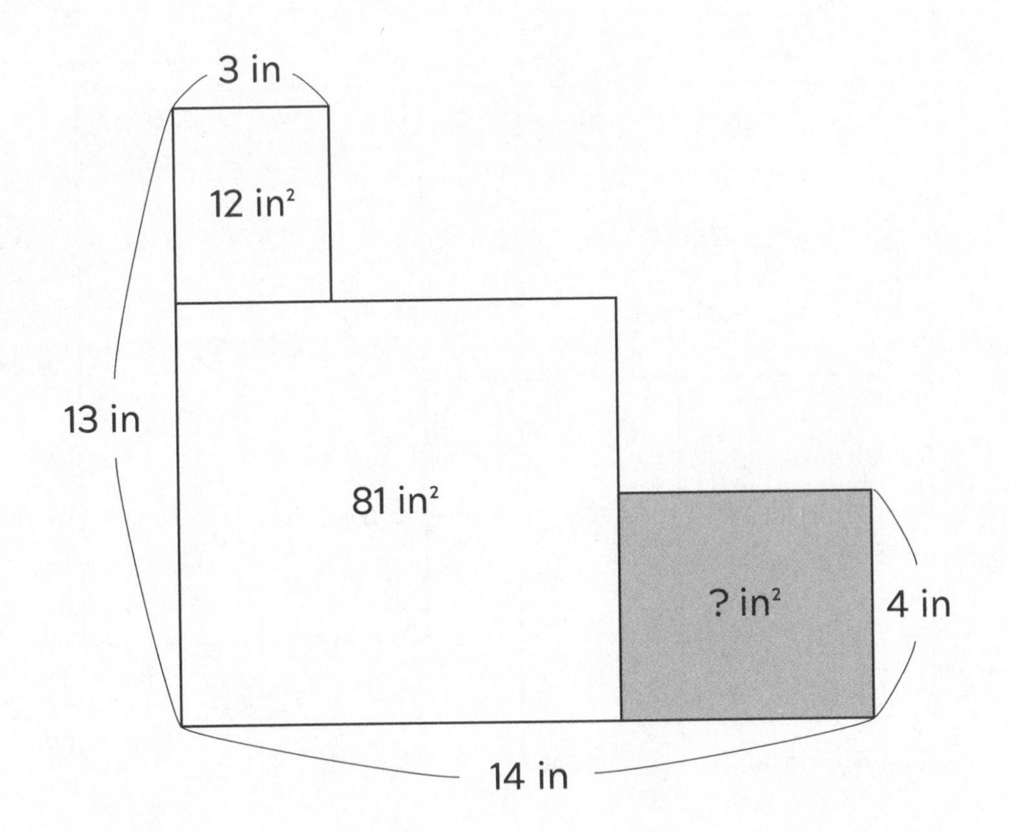
3 in
12 in²
13 in
81 in²
? in²
4 in
14 in

Solution ____________________

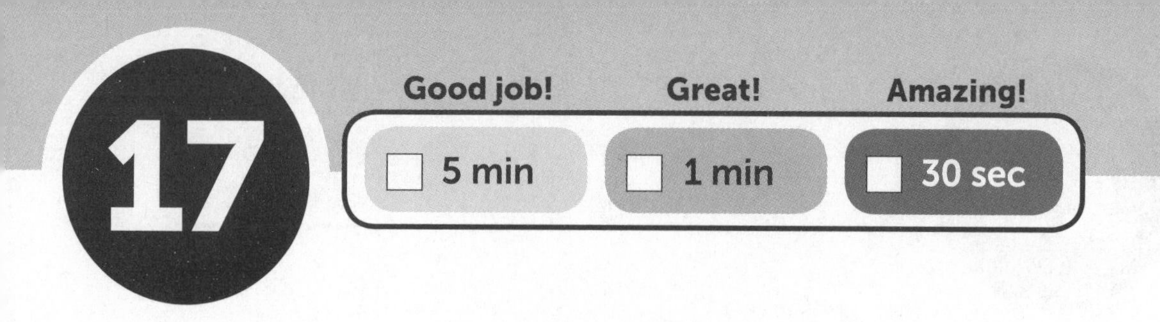
17
Good job!
Great!
Amazing!
5 min
1 min
30 sec

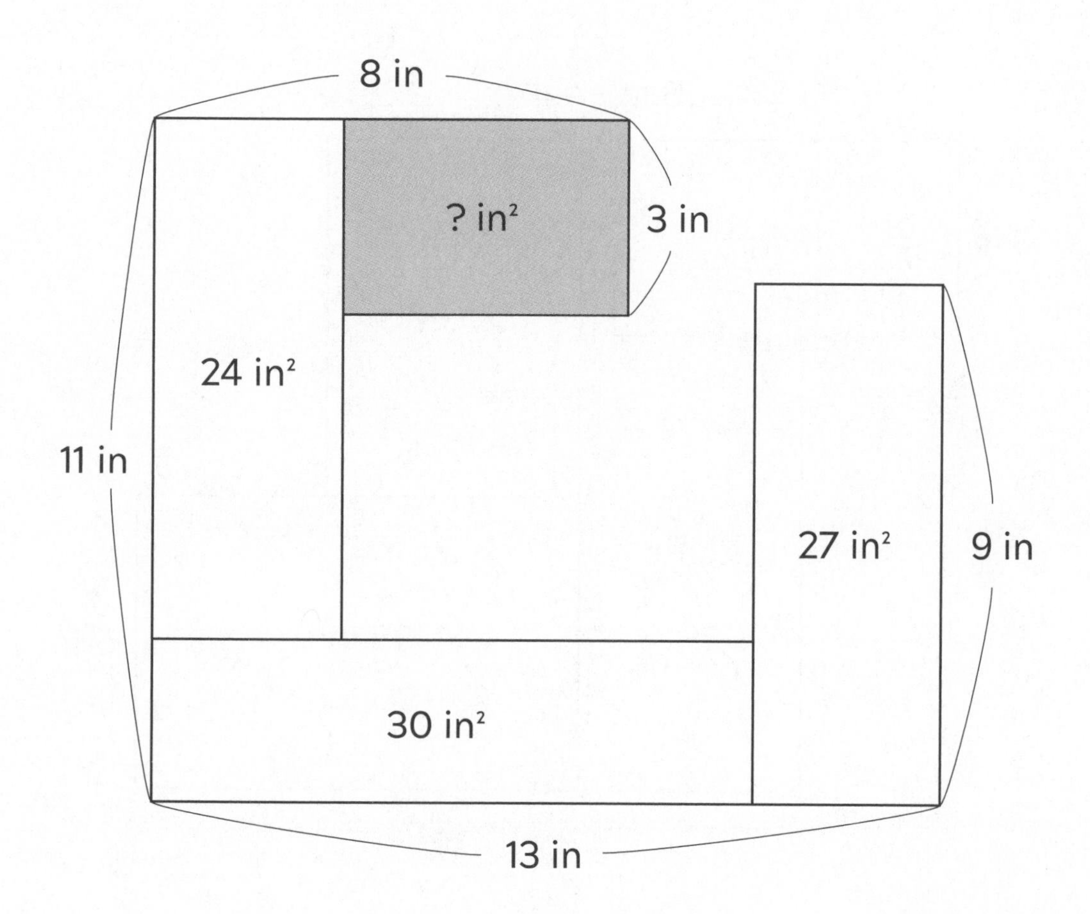
8 in
? in²
3 in
24 in²
11 in
27 in²
9 in
30 in²
13 in

Solution ________________

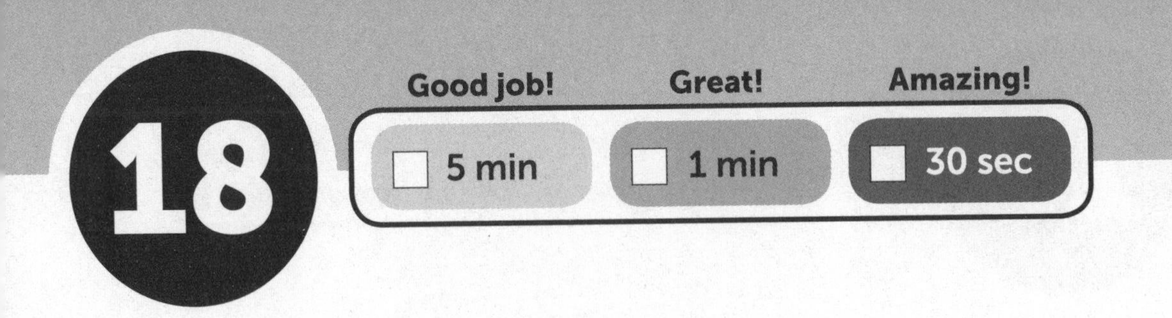
18
Good job!
Great!
Amazing!
5 min
1 min
30 sec

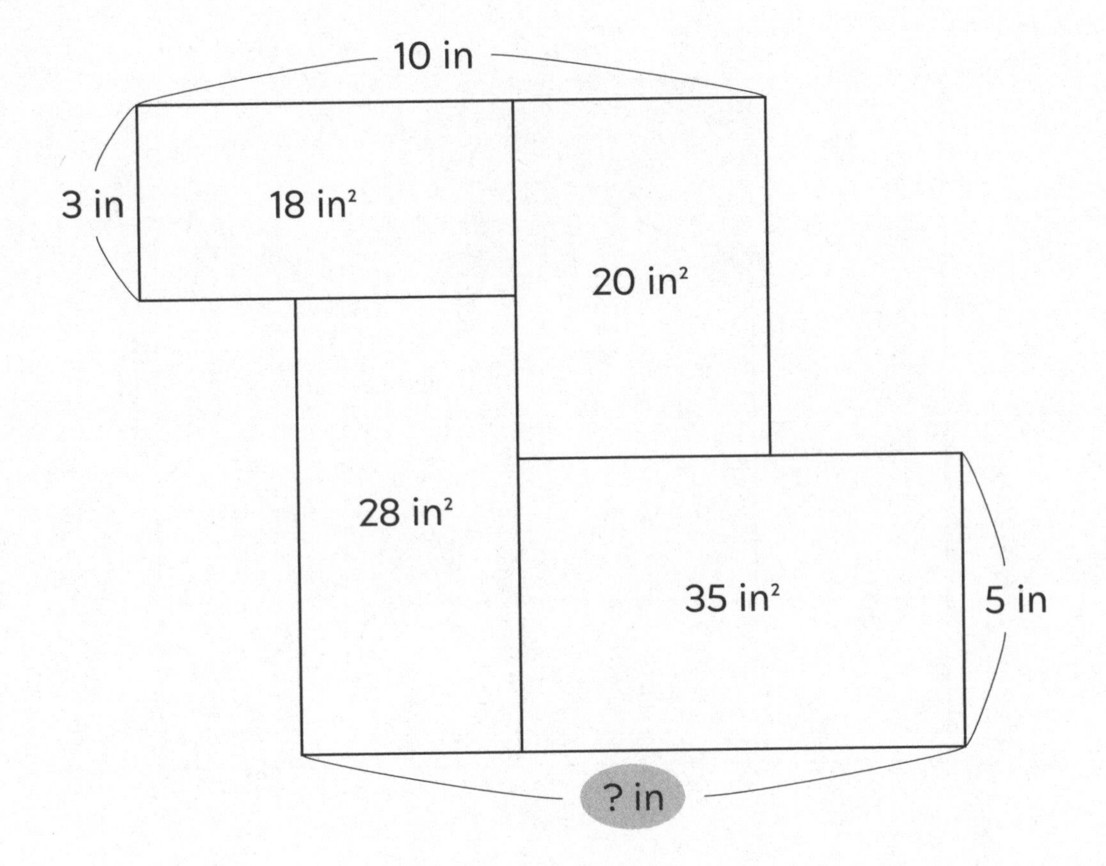
10 in
3 in
18 in²
20 in²
28 in²
35 in²
5 in
? in

Solution ____________

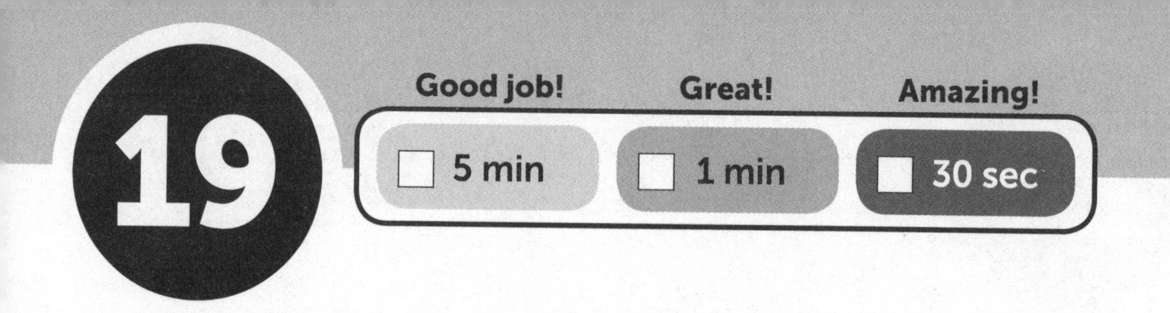

4 in

? in²

36 in²

45 in²

18 in²

27 in²

Solution ____________

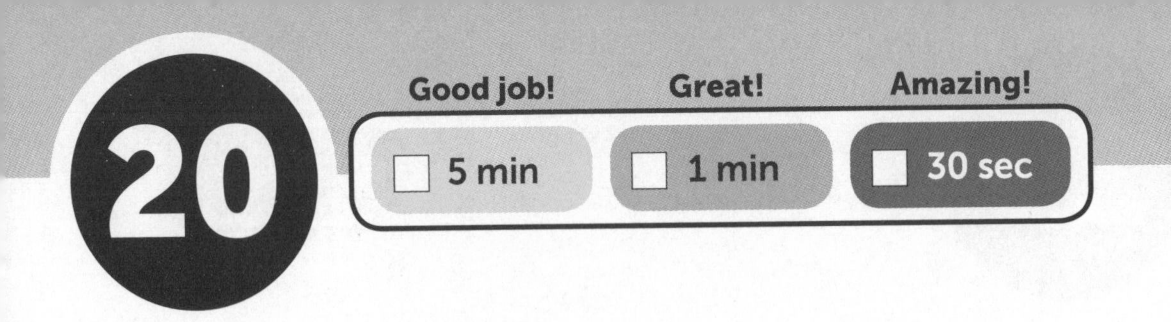
20
Good job!
5 min
Great!
1 min
Amazing!
30 sec

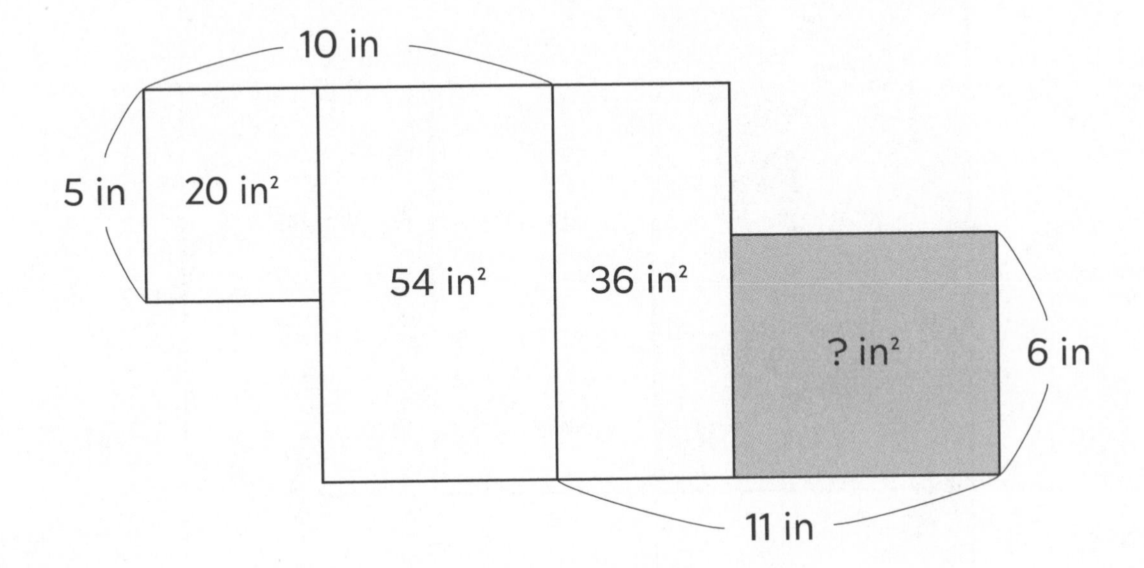
10 in
5 in
20 in²
54 in²
36 in²
? in²
6 in
11 in

Solution ____________

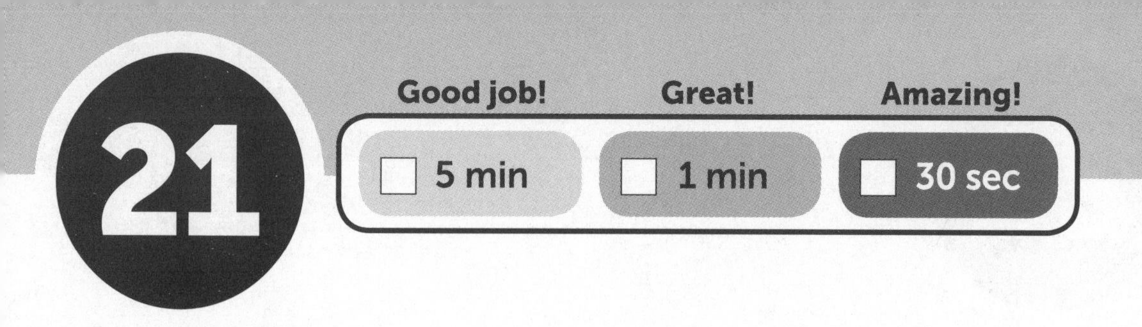
21
Good job!
Great!
Amazing!
5 min
1 min
30 sec

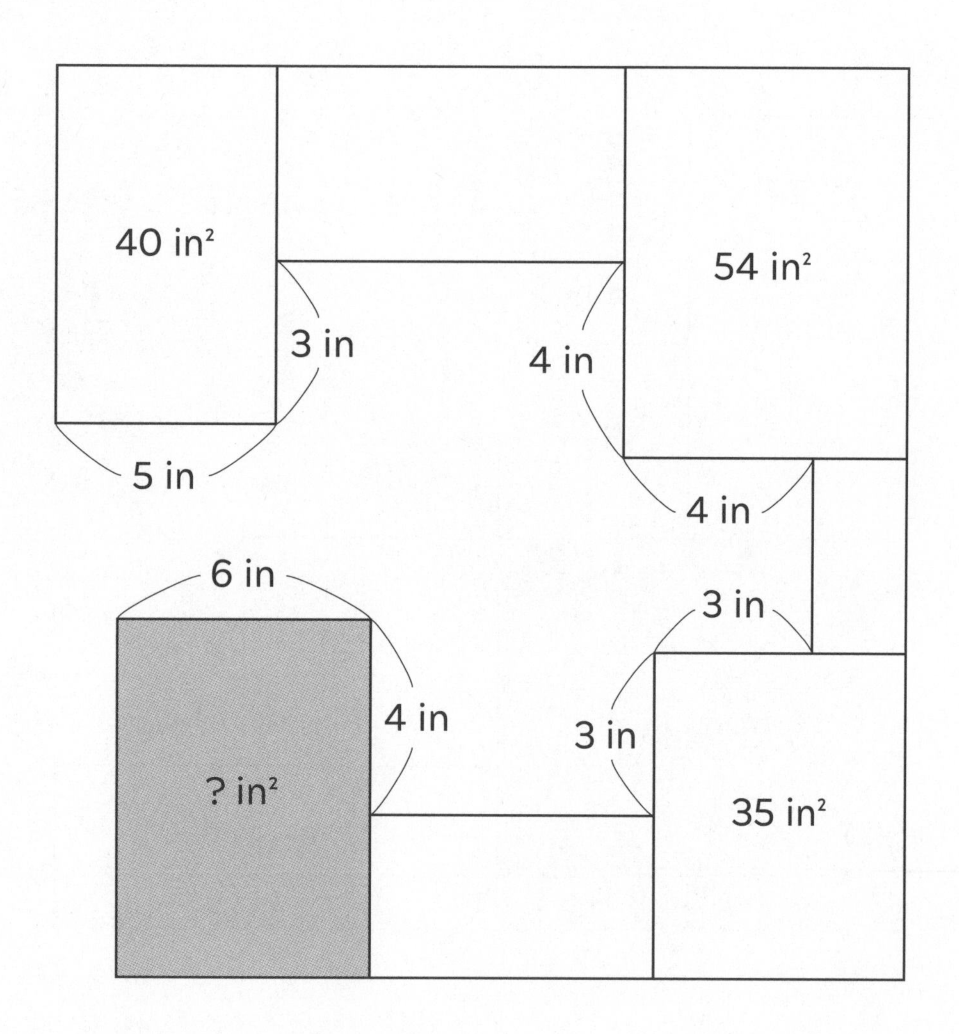
40 in²
54 in²
3 in
4 in
5 in
4 in
6 in
3 in
4 in
3 in
? in²
35 in²

Solution

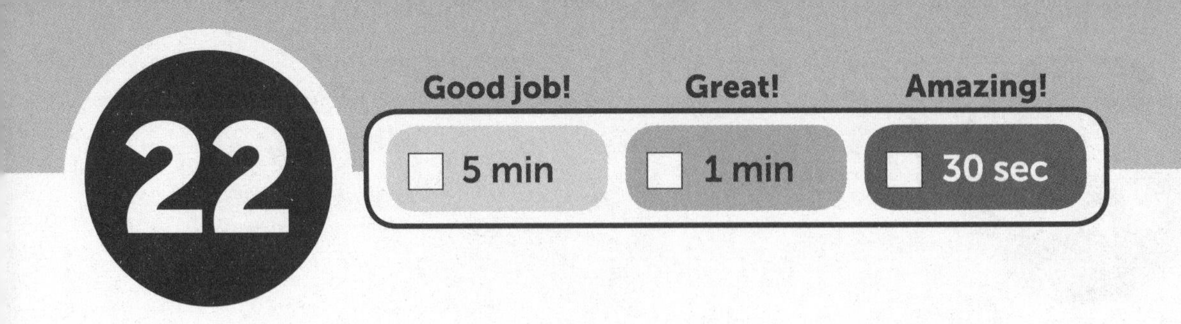
22
Good job!
Great!
Amazing!
5 min
1 min
30 sec

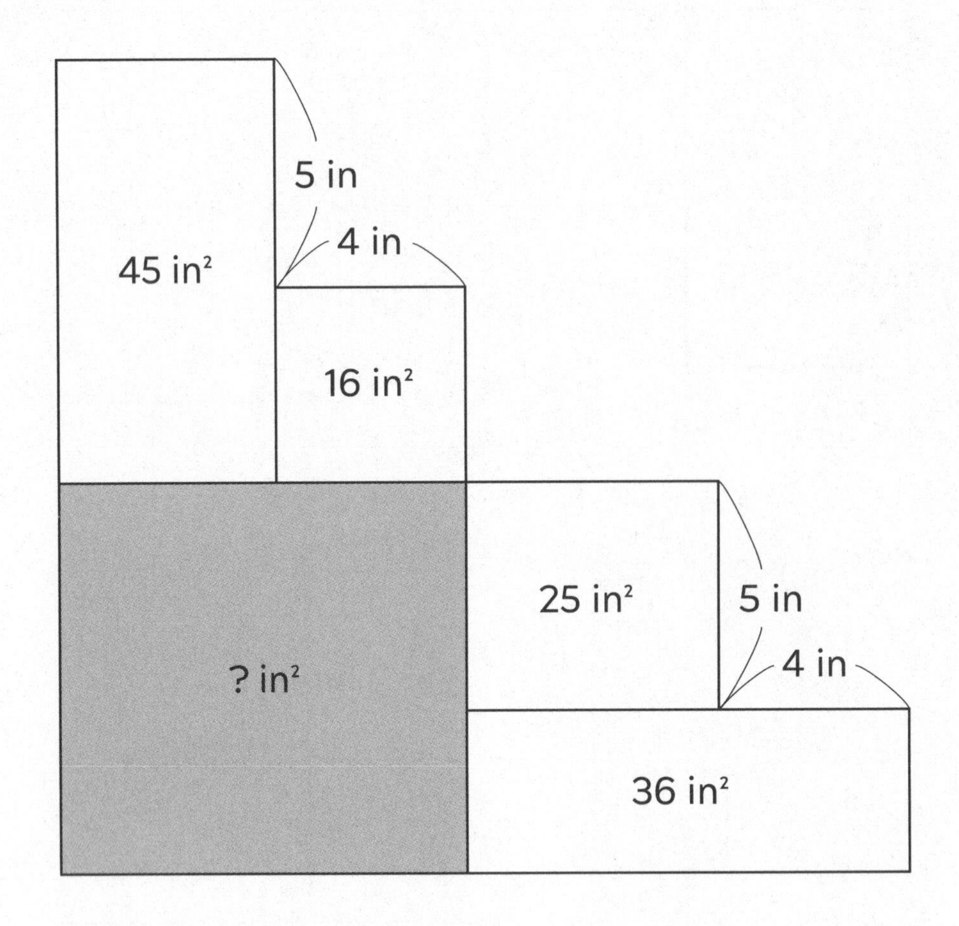
5 in
4 in
45 in²
16 in²
25 in²
5 in
4 in
? in²
36 in²

Solution

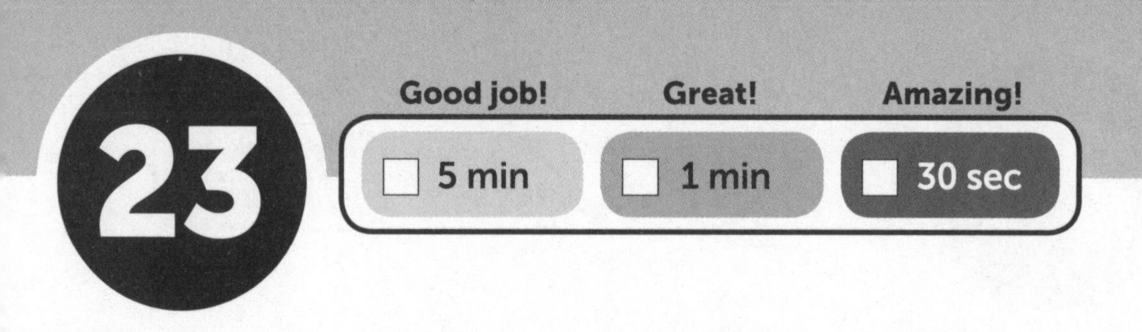

4 in

16 in²

24 in²

28 in²

? in²

20 in²

35 in²

36 in²

30 in²

Solution ________________

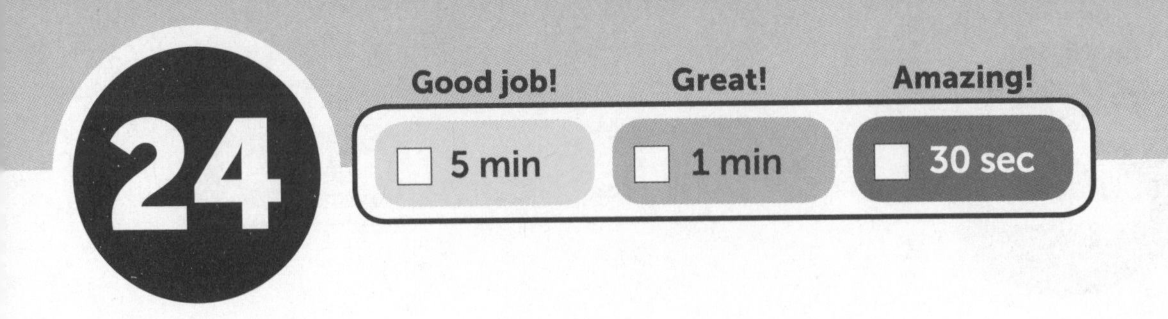
24
Good job!
Great!
Amazing!
5 min
1 min
30 sec

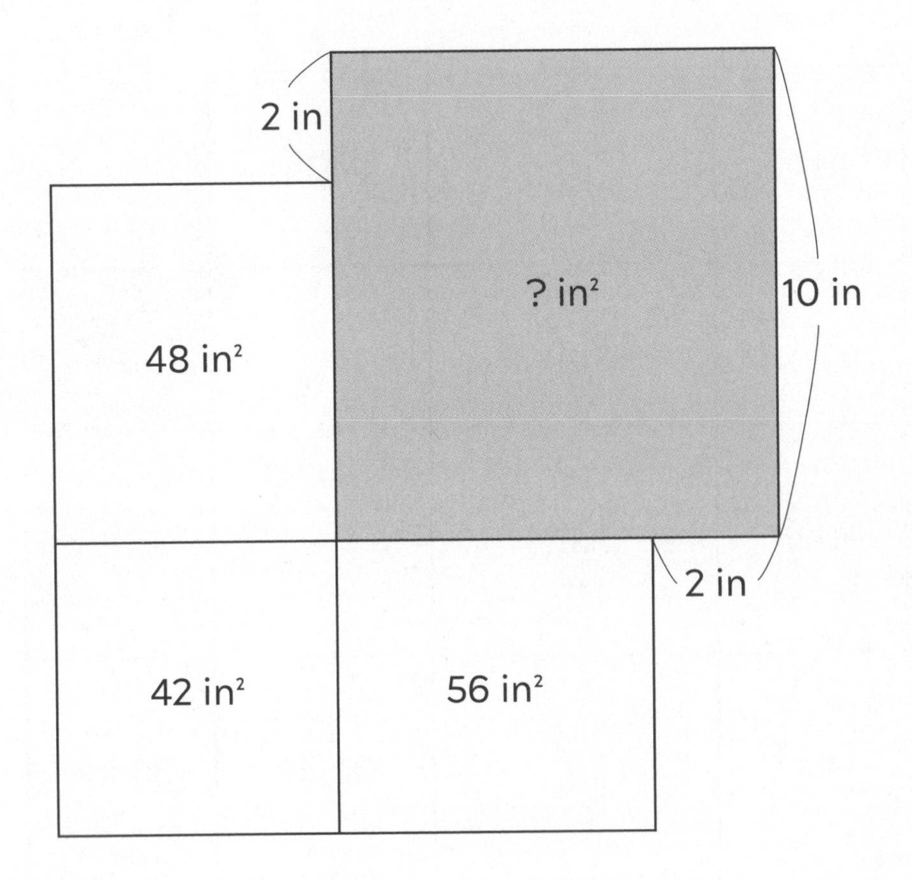
2 in
? in²
10 in
48 in²
2 in
42 in²
56 in²

Solution

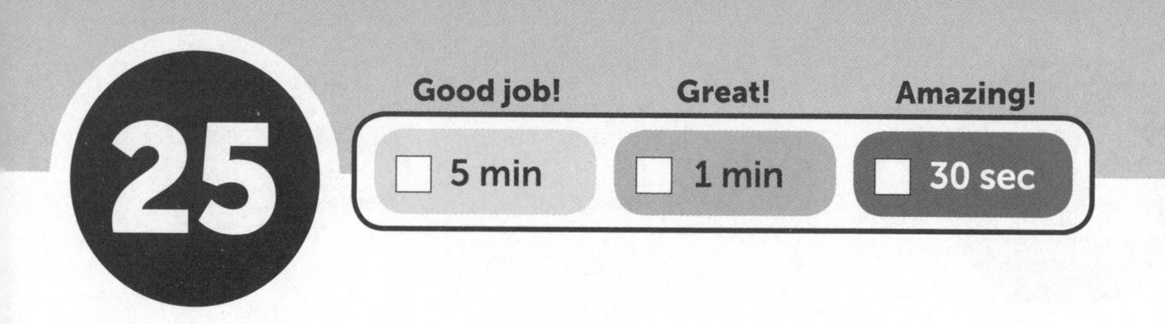
25
Good job!
Great!
Amazing!
5 min
1 min
30 sec

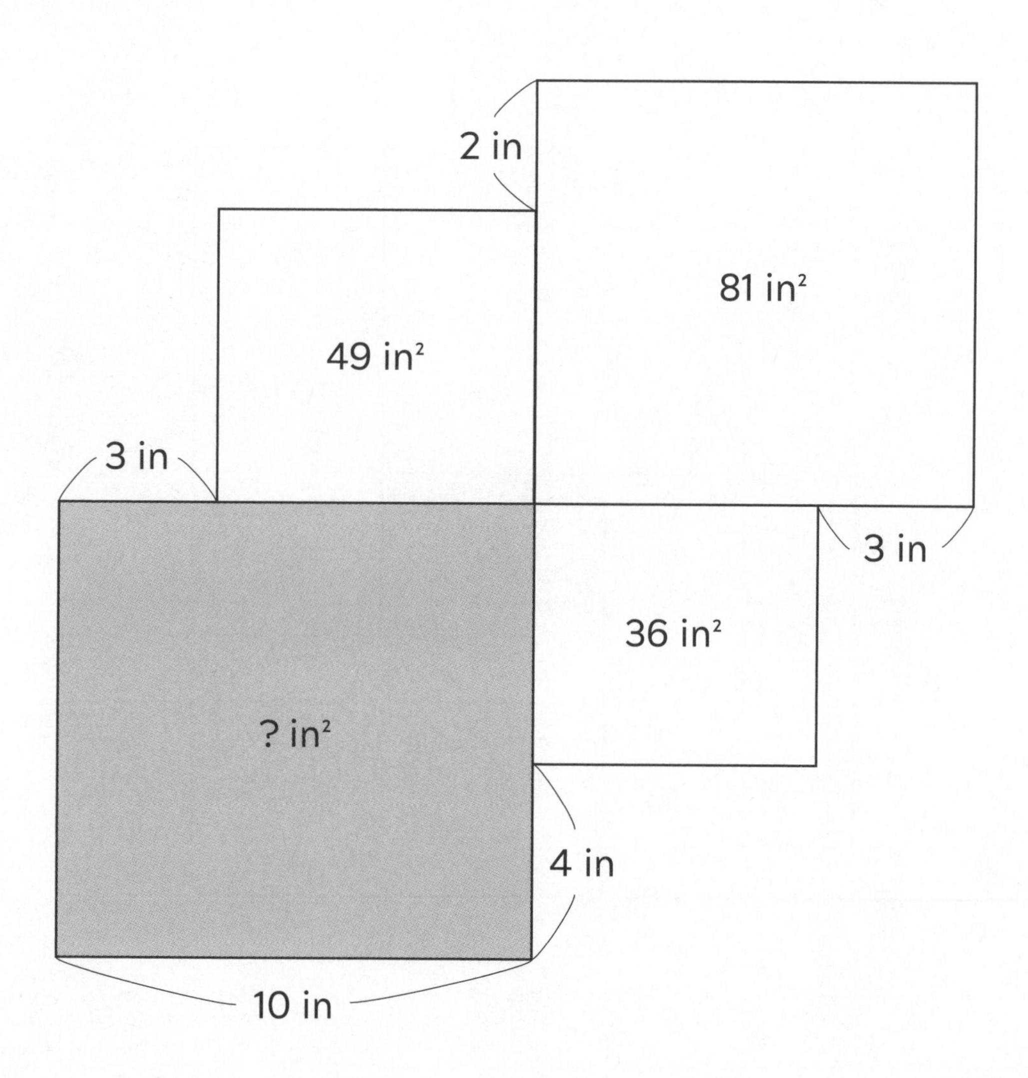
2 in
81 in²
49 in²
3 in
3 in
36 in²
? in²
4 in
10 in

Solution

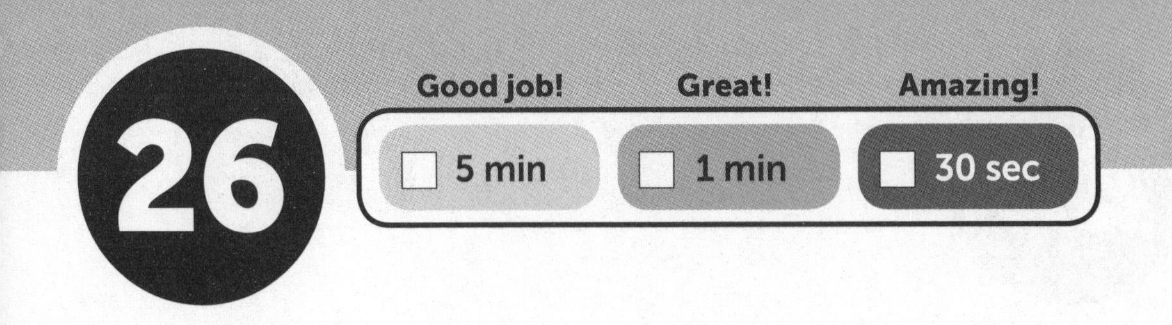
26
Good job!
Great!
Amazing!
5 min
1 min
30 sec

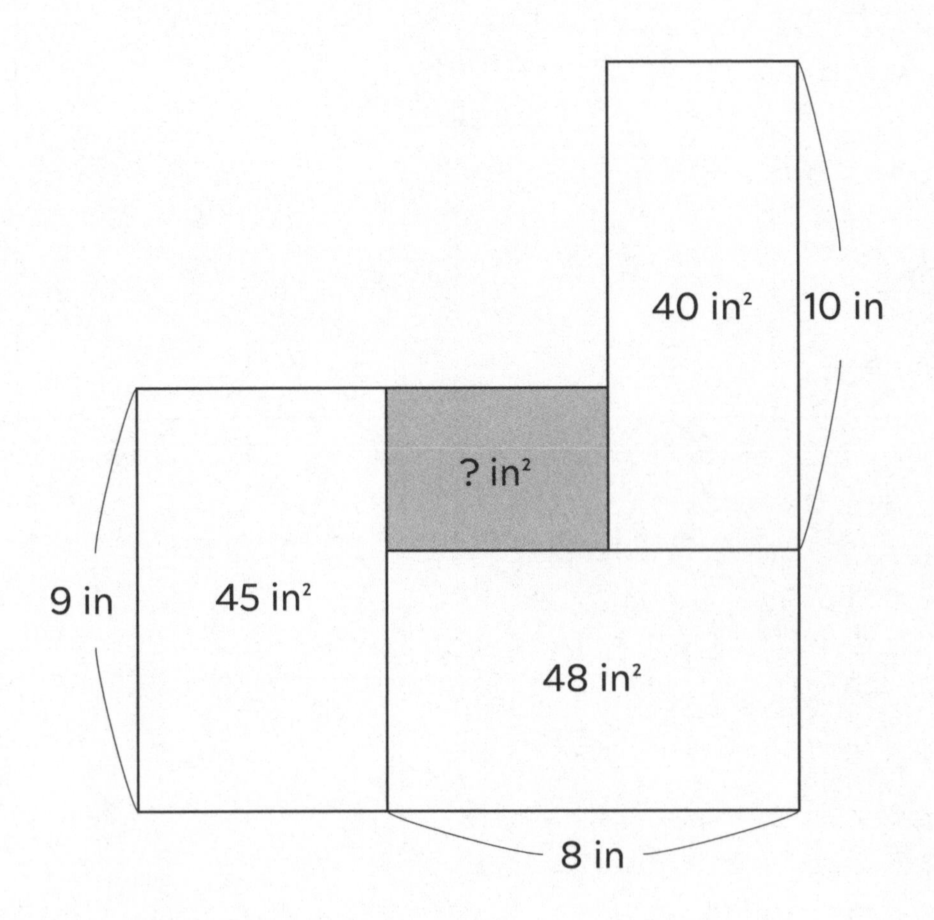
40 in²
10 in
? in²
9 in
45 in²
48 in²
8 in

Solution ______

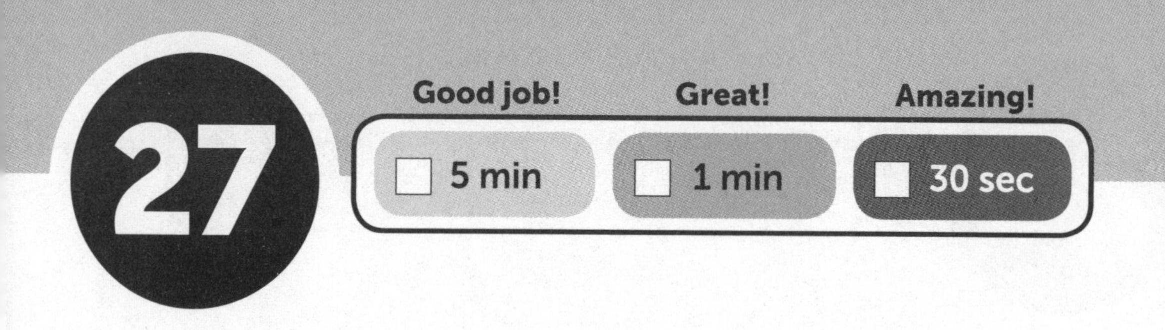
27
Good job!
Great!
Amazing!
5 min
1 min
30 sec

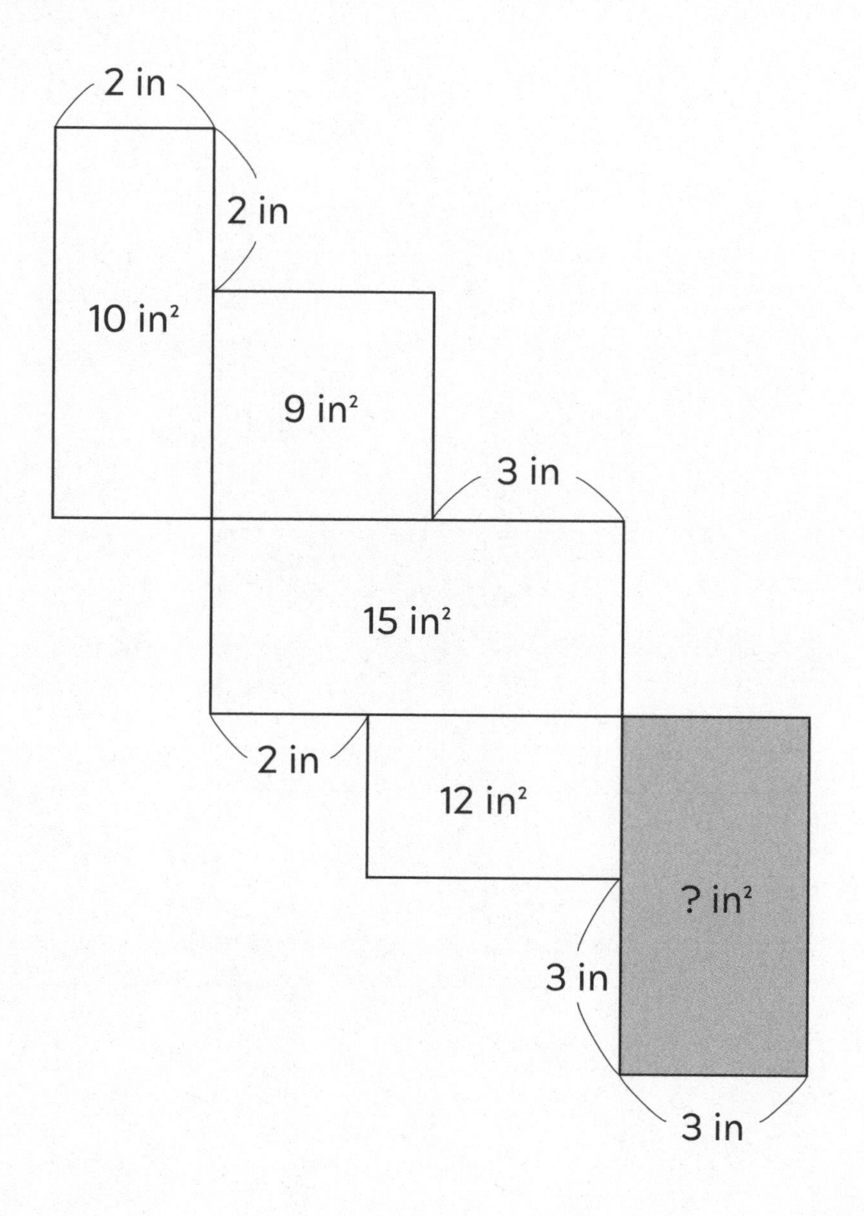
2 in
2 in
10 in²
9 in²
3 in
15 in²
2 in
12 in²
? in²
3 in
3 in

Solution

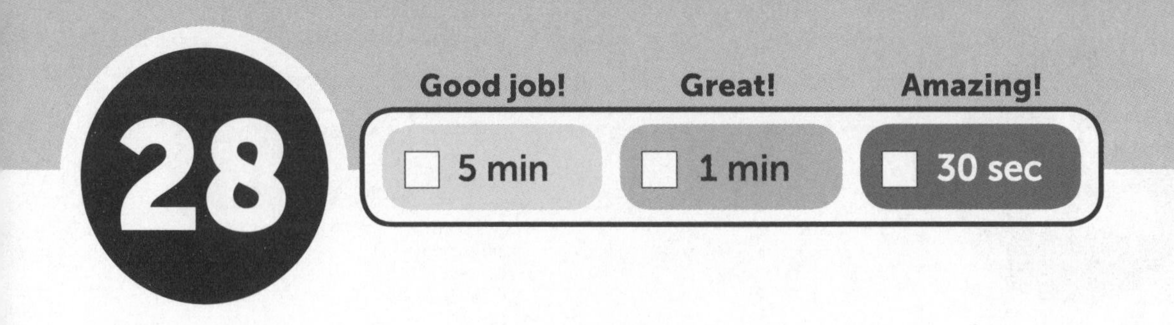
28
Good job!
Great!
Amazing!
5 min
1 min
30 sec

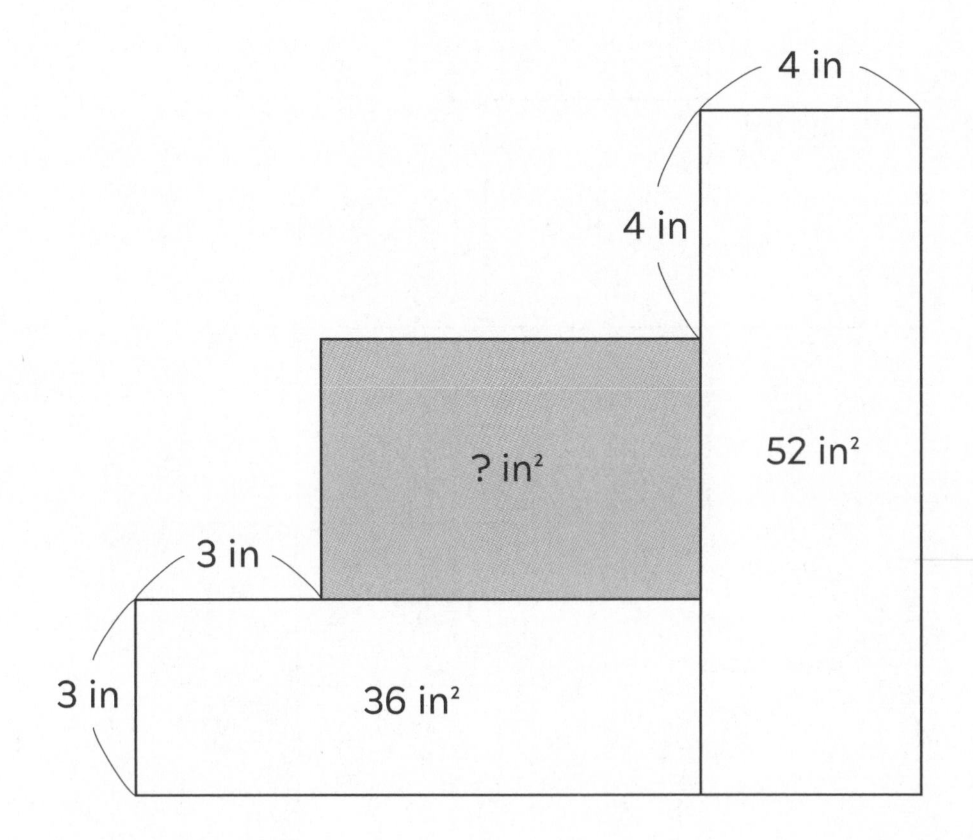
4 in
4 in
? in²
52 in²
3 in
3 in
36 in²

Solution ______________________

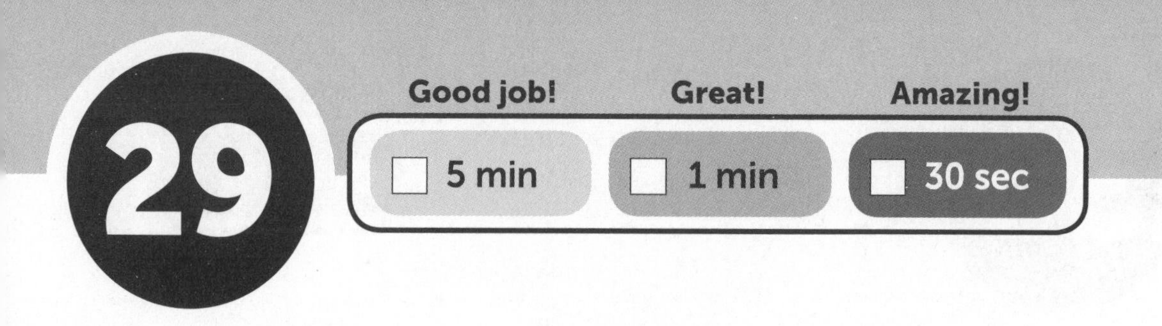
29
Good job!
Great!
Amazing!
5 min
1 min
30 sec

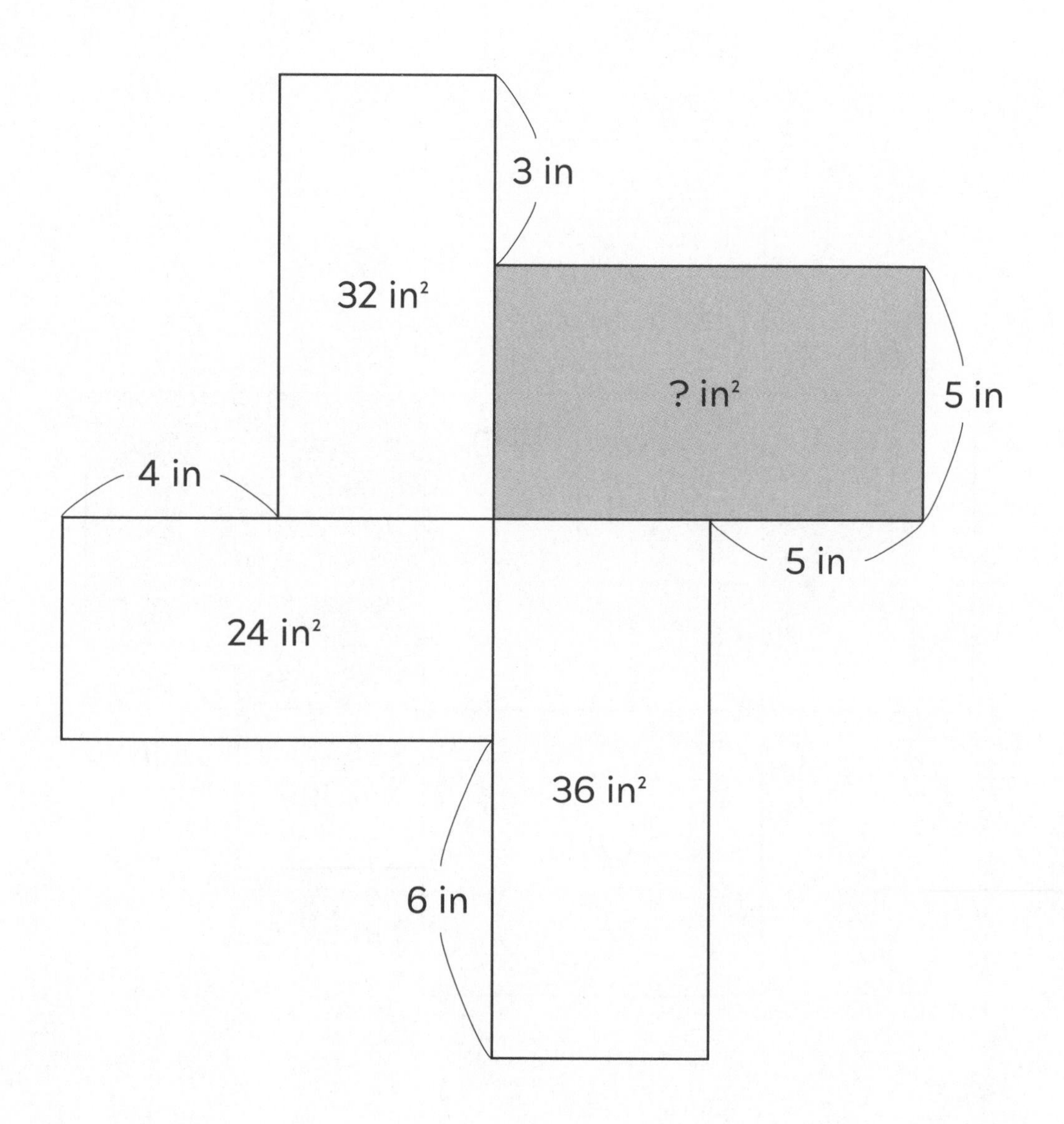
3 in
32 in²
? in²
5 in
4 in
5 in
24 in²
36 in²
6 in

Solution ____________

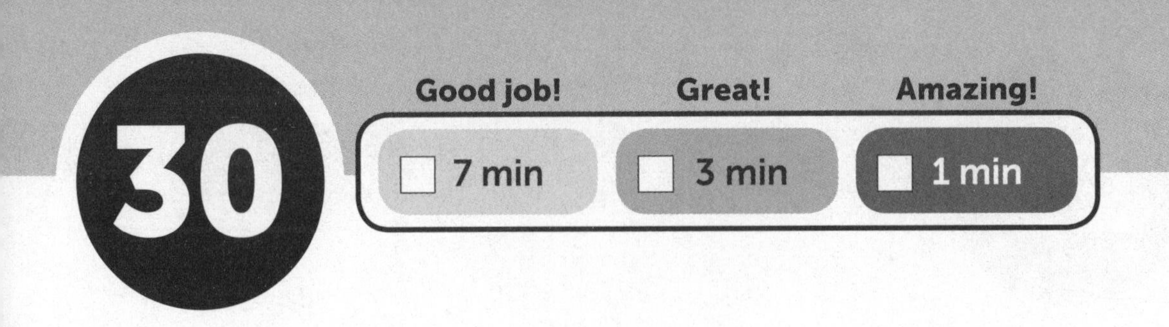
30
Good job!
Great!
Amazing!
7 min
3 min
1 min

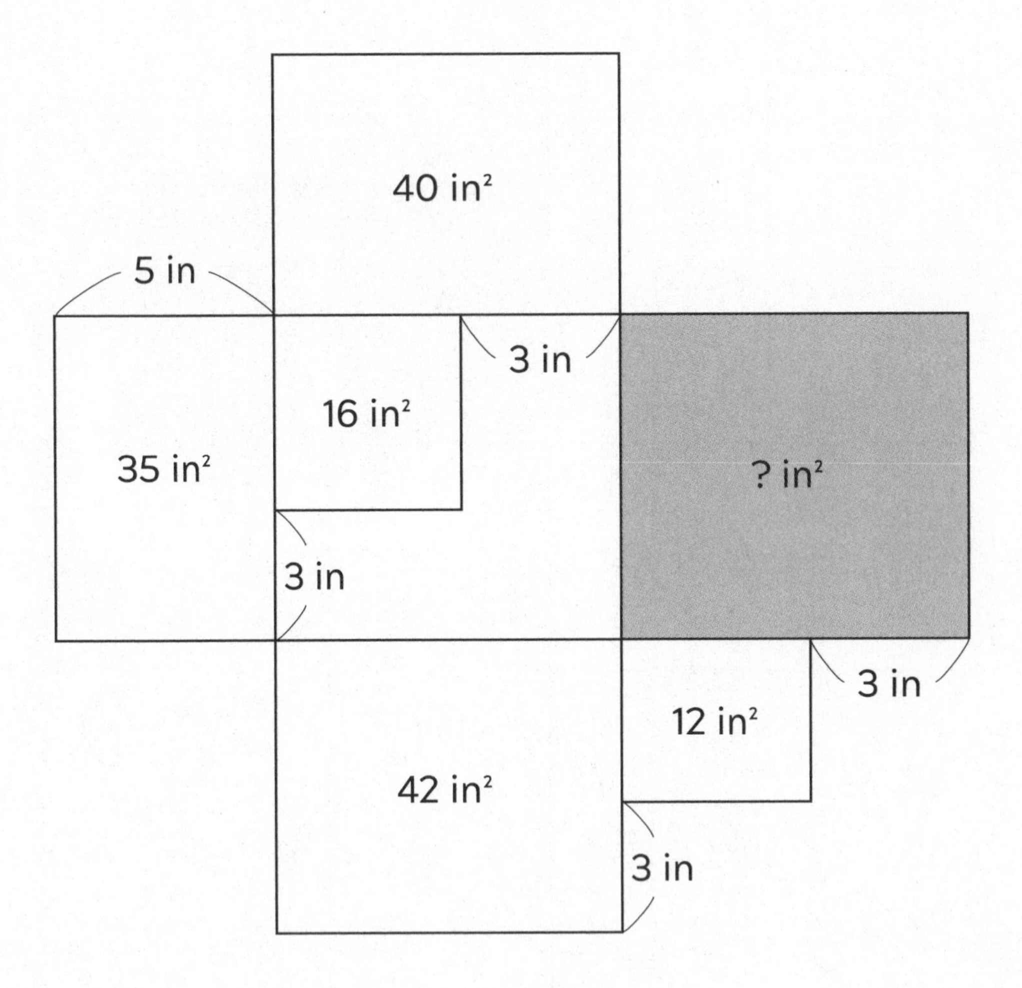
40 in²
5 in
3 in
16 in²
35 in²
? in²
3 in
3 in
12 in²
42 in²
3 in

Solution ________________

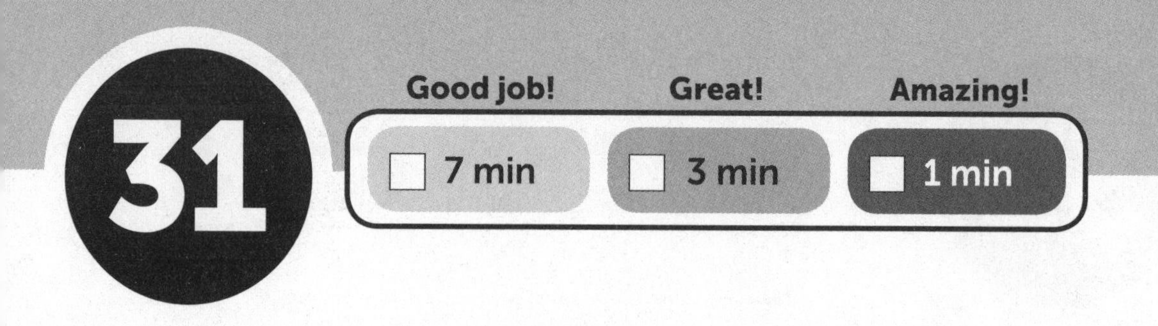
31
Good job!
Great!
Amazing!
7 min
3 min
1 min

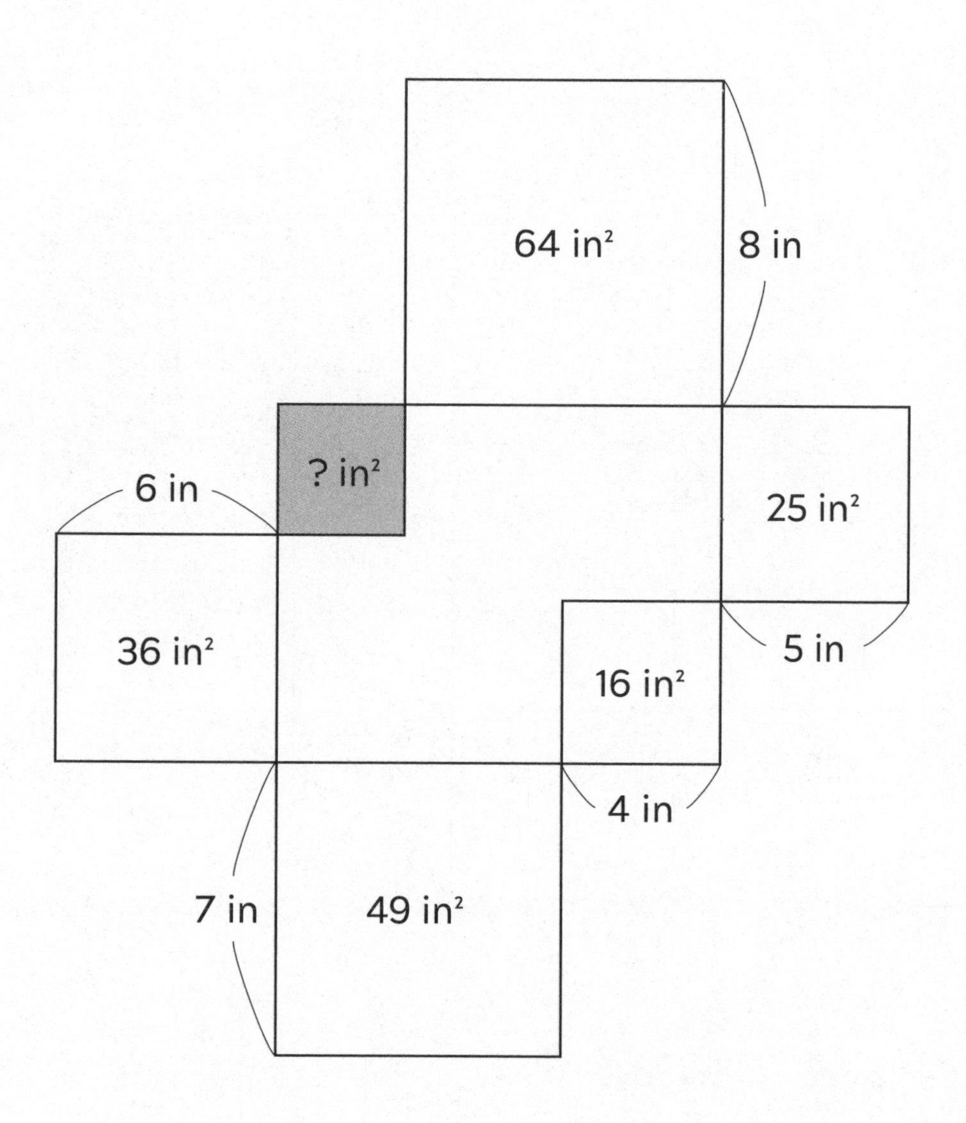
64 in²
8 in
? in²
6 in
25 in²
36 in²
5 in
16 in²
4 in
7 in
49 in²

Solution

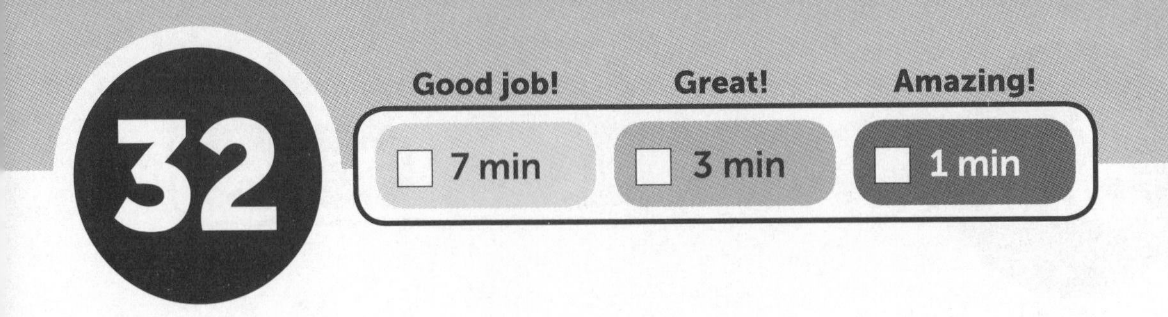
32
Good job!
Great!
Amazing!
7 min
3 min
1 min

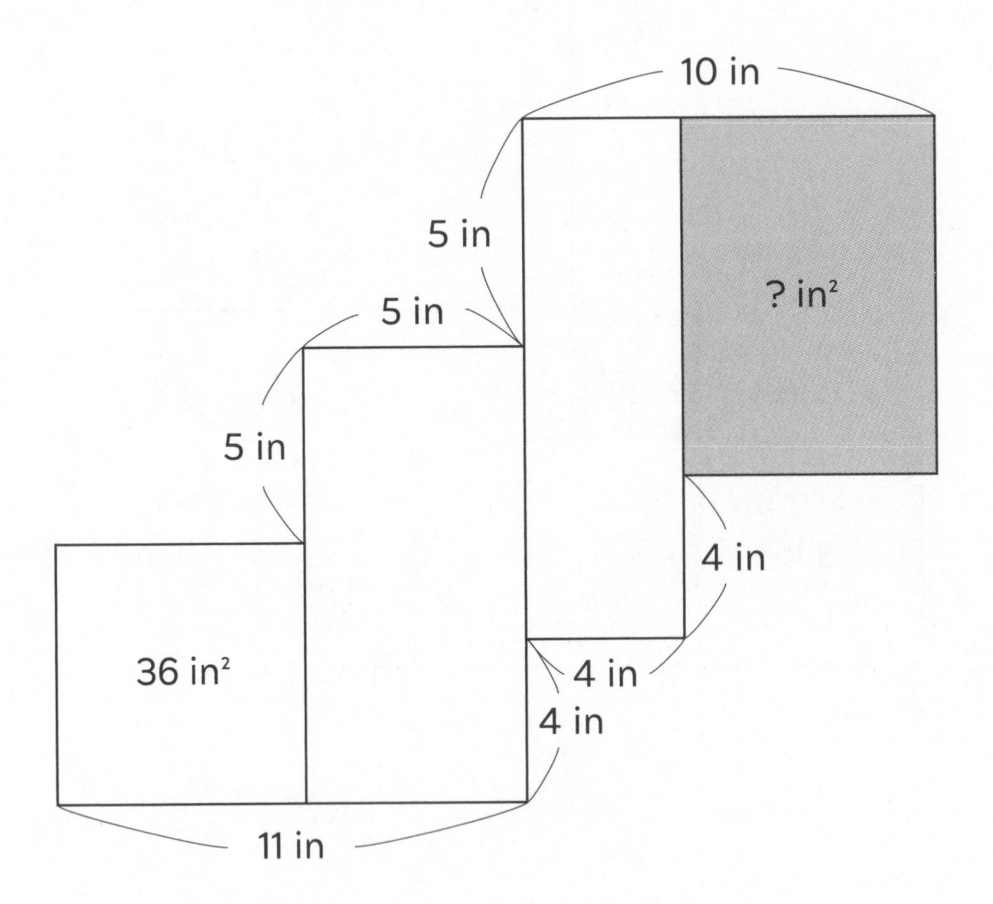
10 in
5 in
5 in
5 in
? in²
4 in
36 in²
4 in
4 in
11 in

Solution

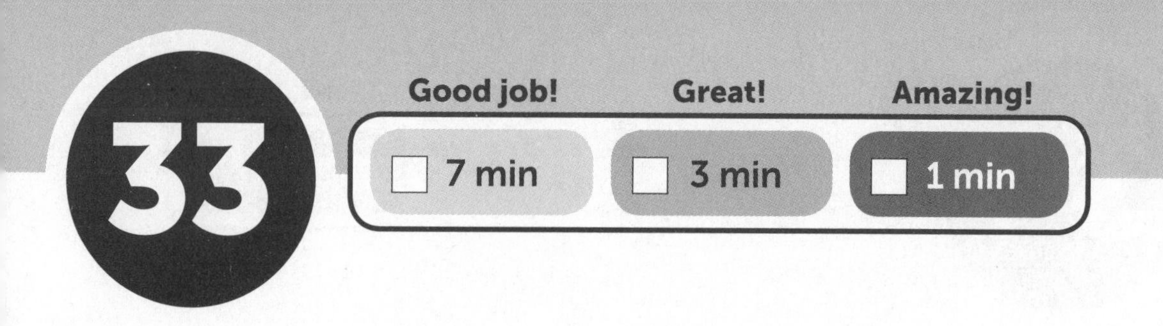
33
Good job!
Great!
Amazing!
7 min
3 min
1 min

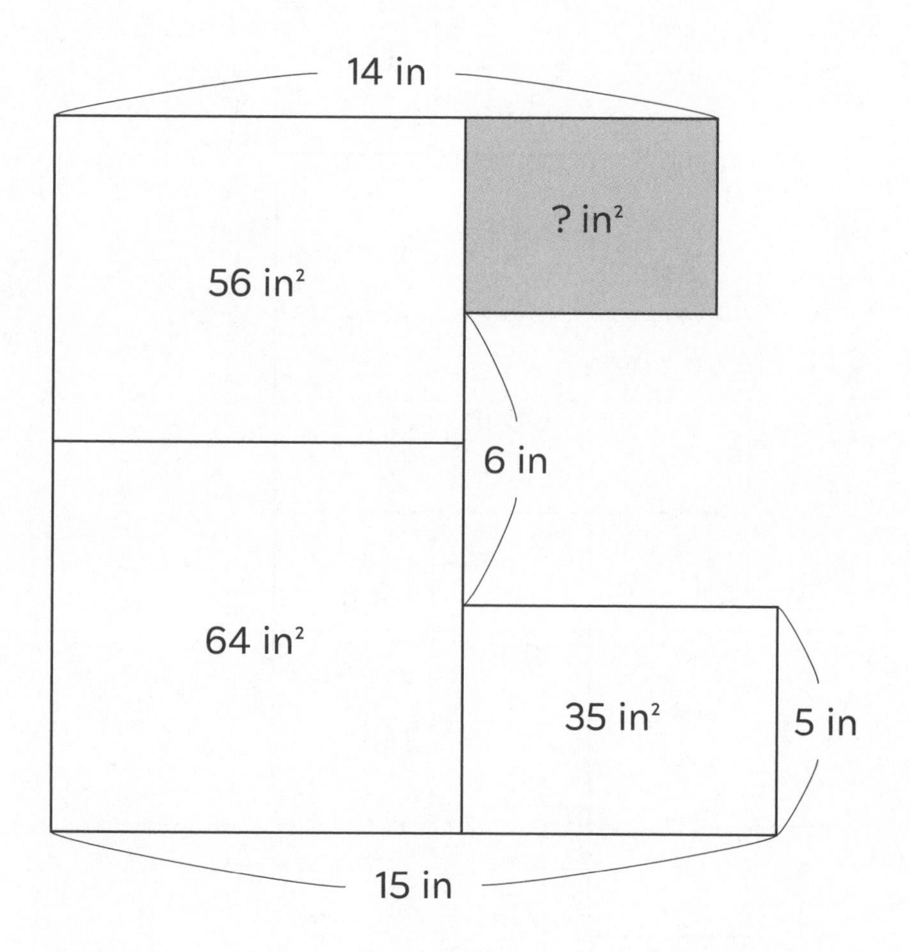
14 in
? in²
56 in²
6 in
64 in²
35 in²
5 in
15 in

Solution

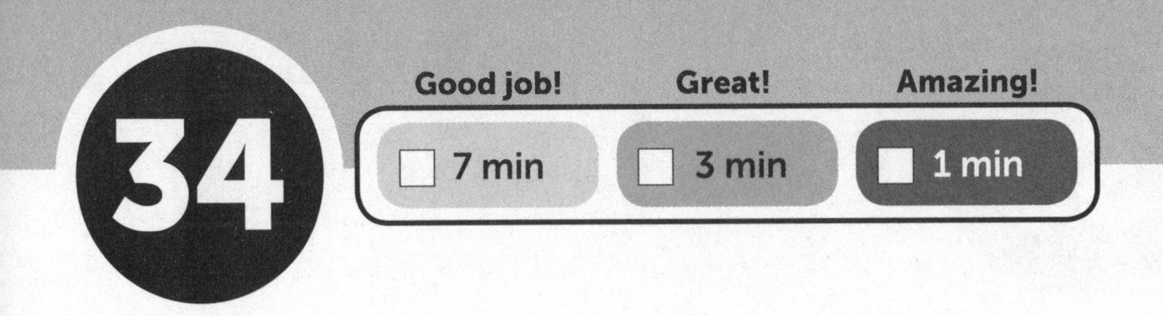
34
Good job!
Great!
Amazing!
7 min
3 min
1 min

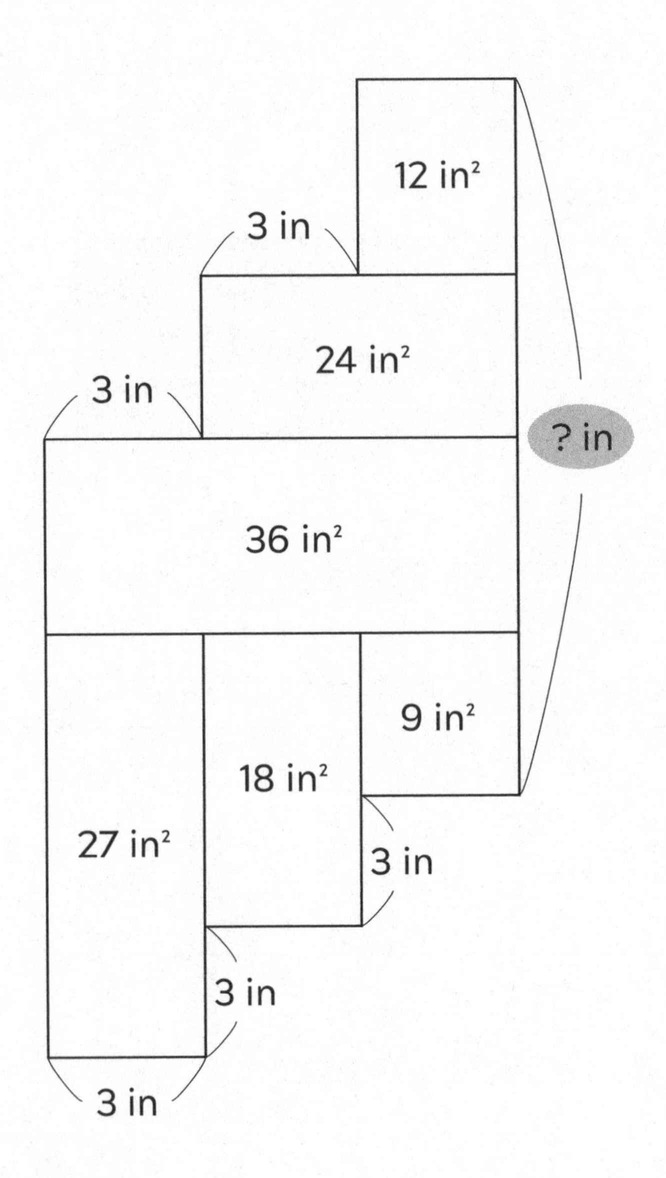
12 in²
3 in
24 in²
3 in
? in
36 in²
9 in²
18 in²
27 in²
3 in
3 in
3 in

Solution

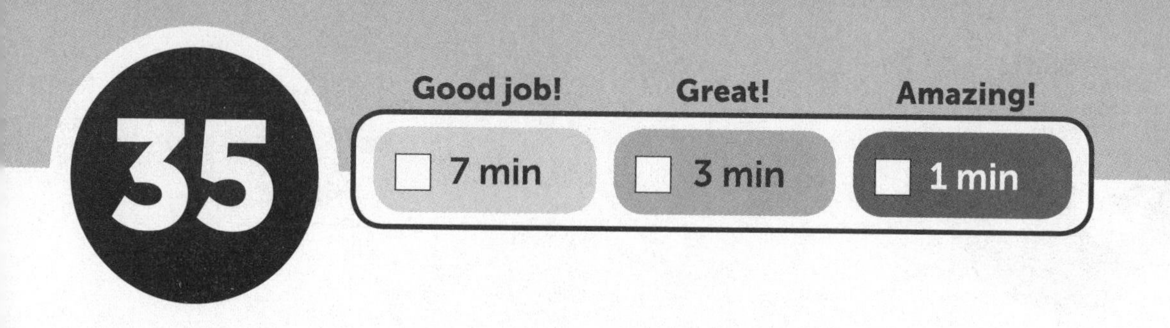
35
Good job!
7 min
Great!
3 min
Amazing!
1 min

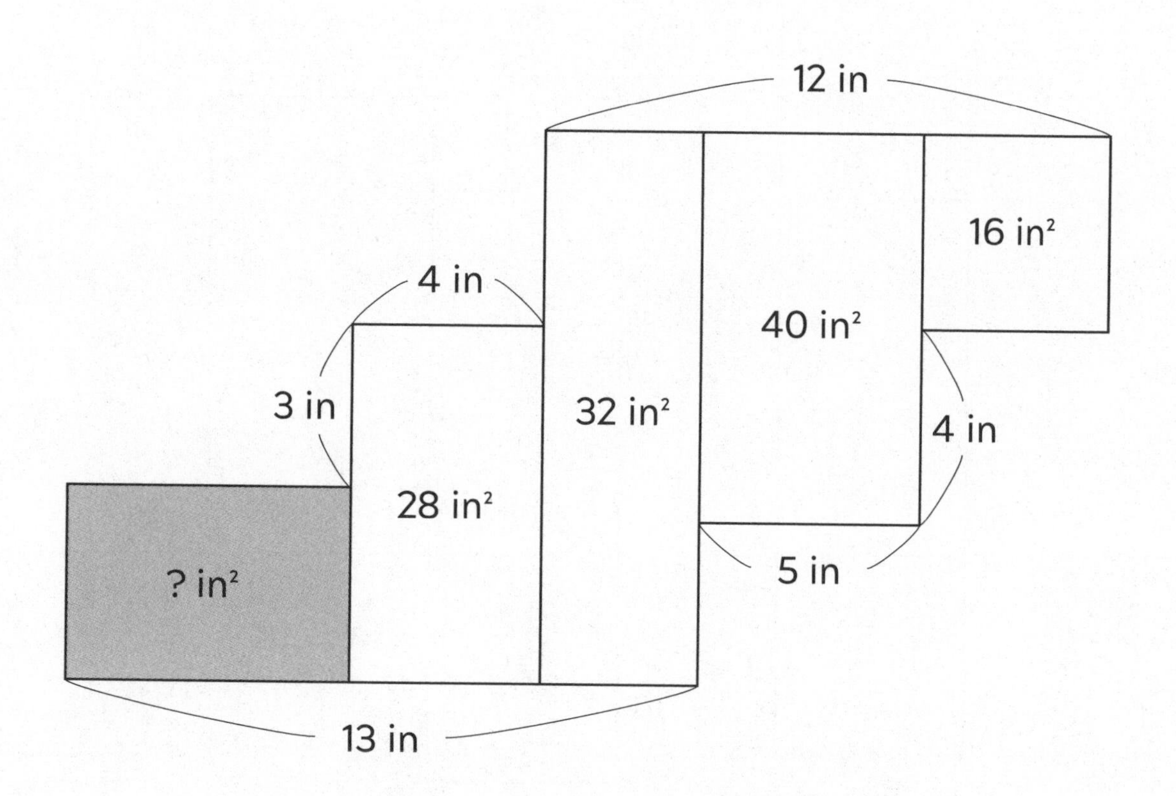
12 in
16 in²
4 in
40 in²
3 in
32 in²
4 in
28 in²
? in²
5 in
13 in

Solution

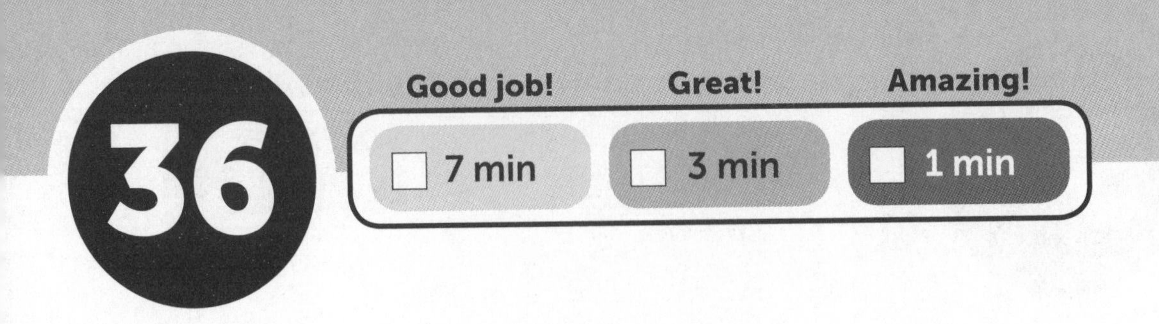
36
Good job!
Great!
Amazing!
7 min
3 min
1 min

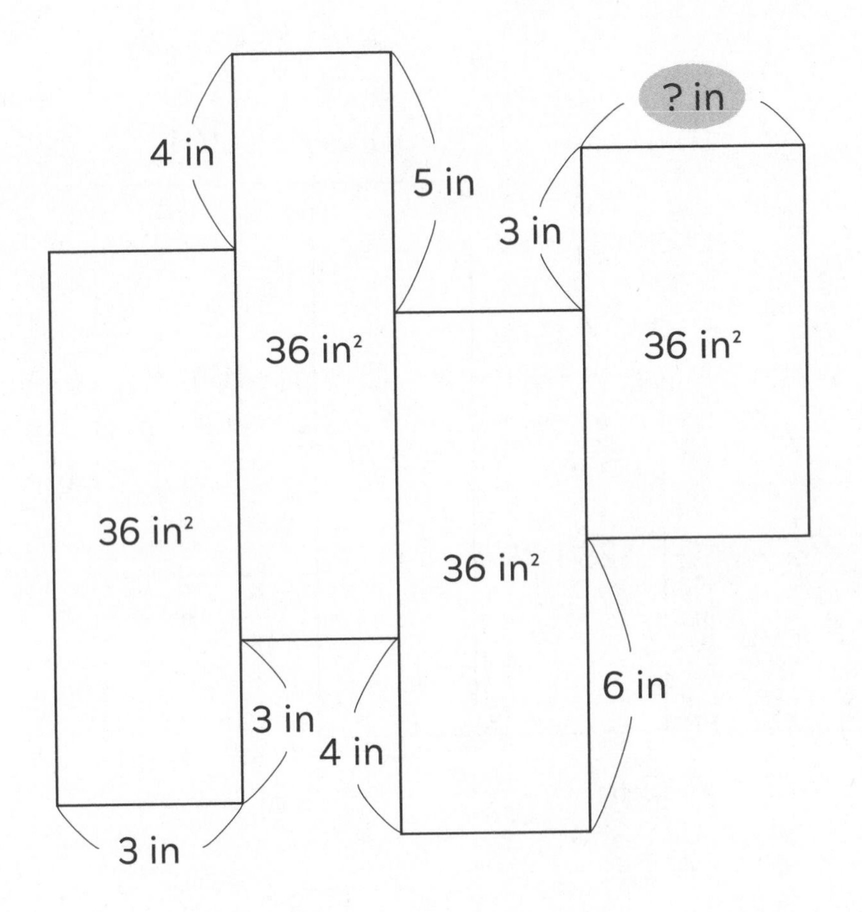
4 in
5 in
? in
3 in
36 in²
36 in²
36 in²
36 in²
6 in
3 in
4 in
3 in

Solution ______________________

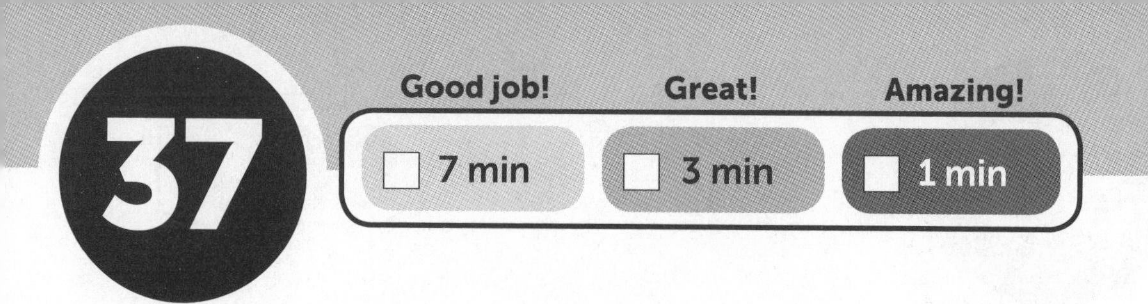
37
Good job!
Great!
Amazing!
7 min
3 min
1 min

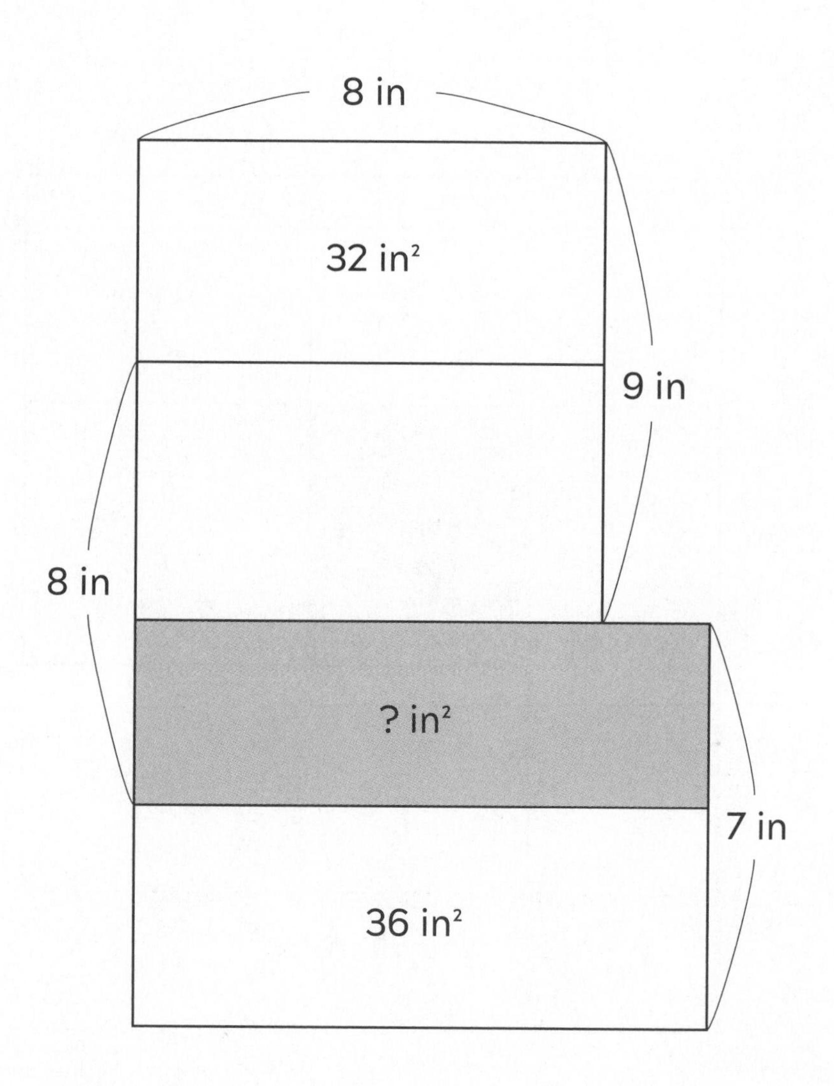
8 in
32 in²
9 in
8 in
? in²
7 in
36 in²

Solution

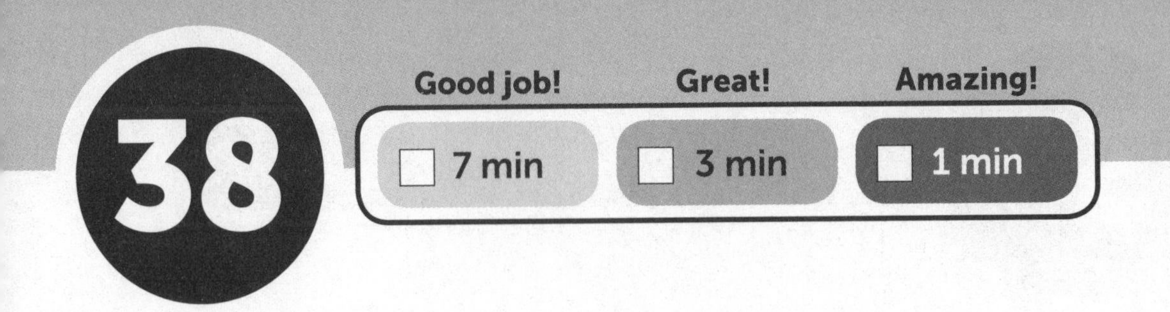
38
Good job!
Great!
Amazing!
7 min
3 min
1 min

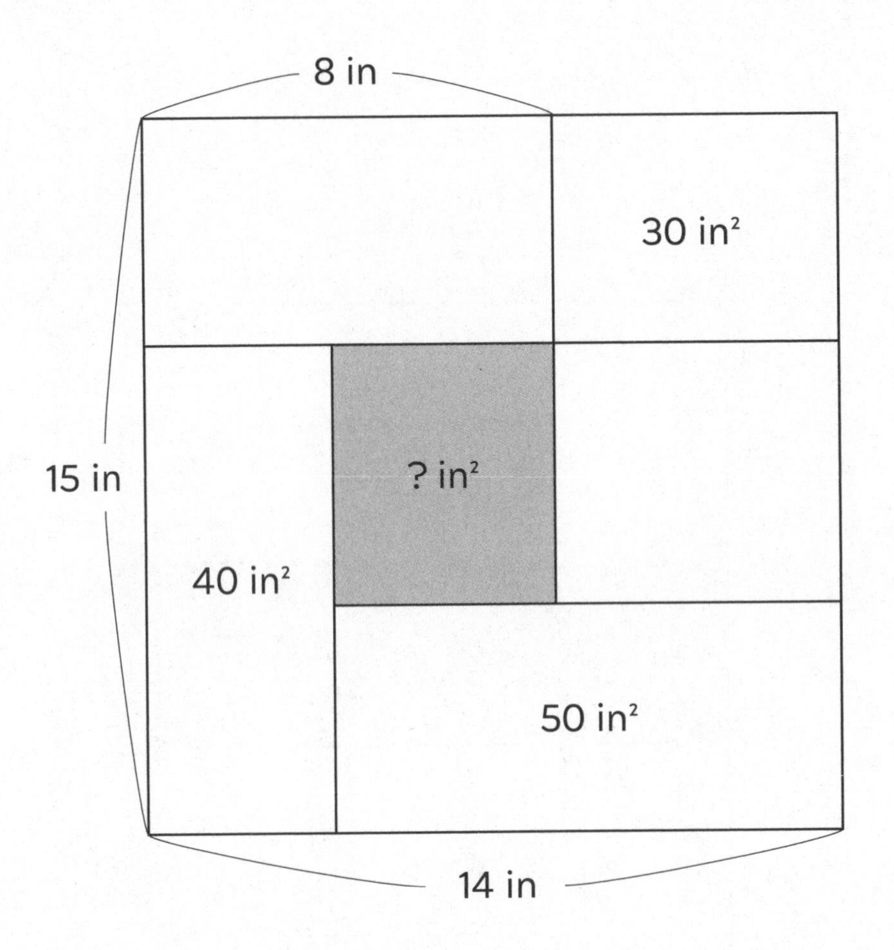
8 in
30 in²
15 in
? in²
40 in²
50 in²
14 in

Solution ________________

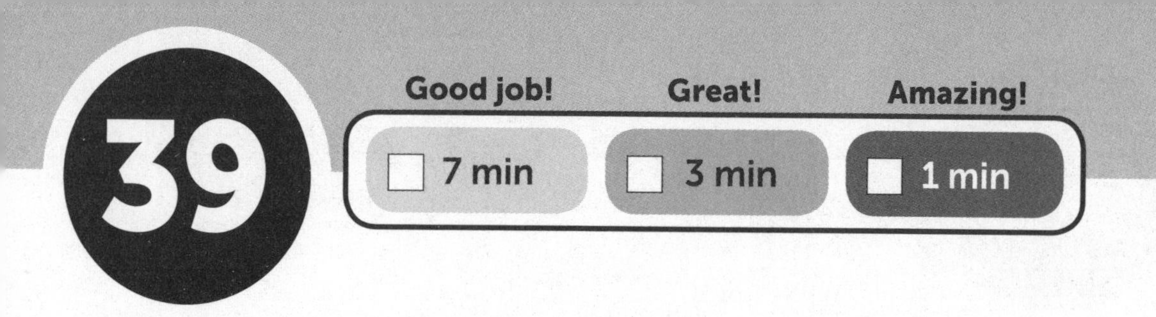
39
Good job!
Great!
Amazing!
7 min
3 min
1 min

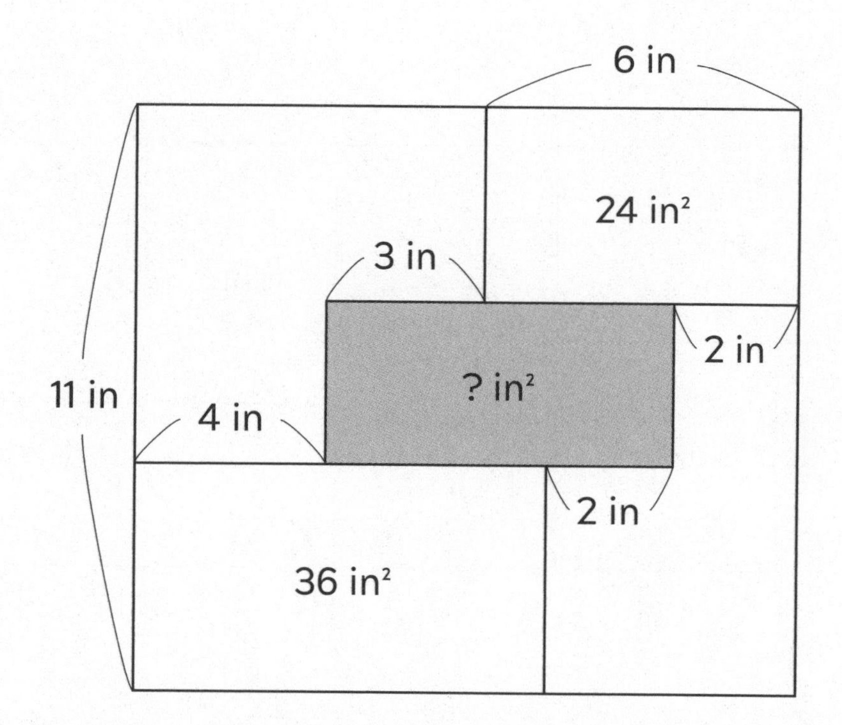
6 in
24 in²
3 in
2 in
11 in
? in²
4 in
2 in
36 in²

Solution ______________________

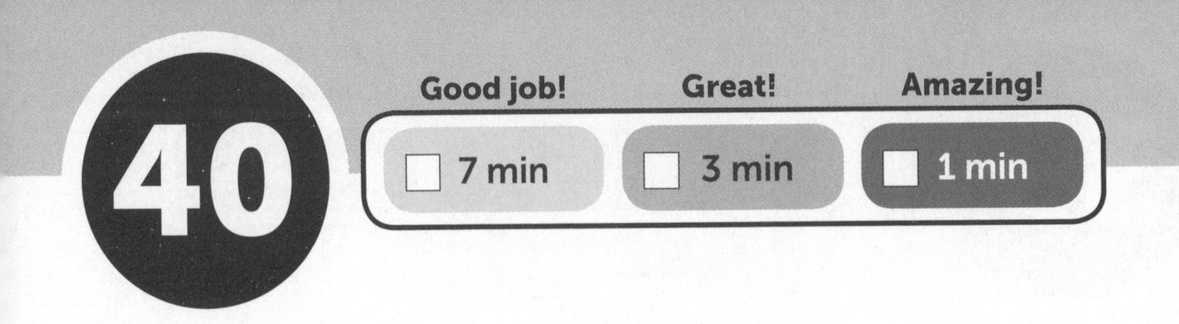
40
Good job!
Great!
Amazing!
7 min
3 min
1 min

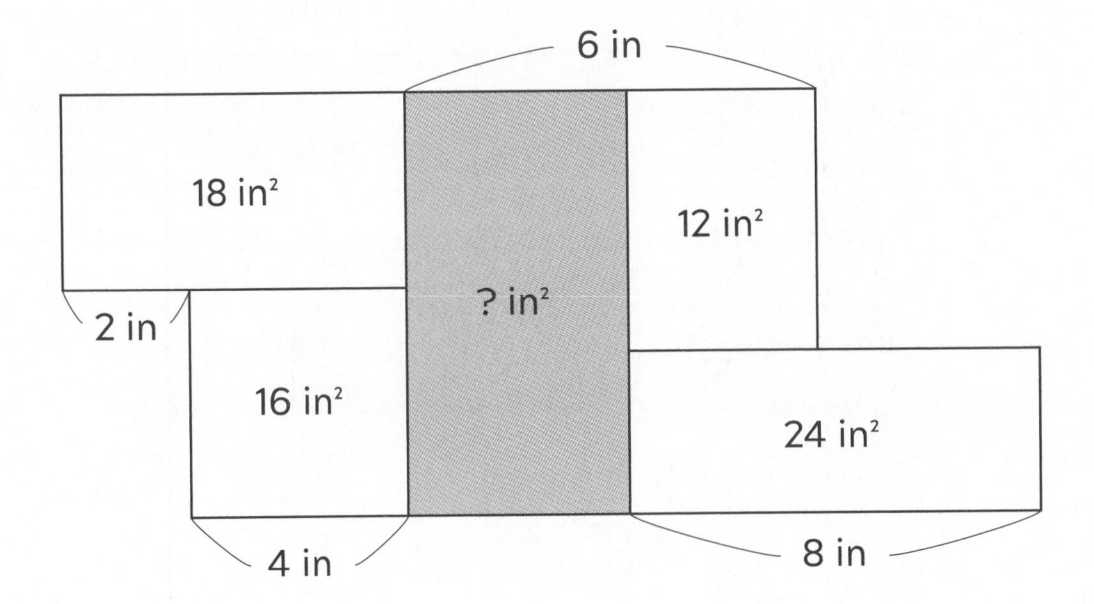
6 in
18 in²
12 in²
? in²
2 in
16 in²
24 in²
4 in
8 in

Solution ______

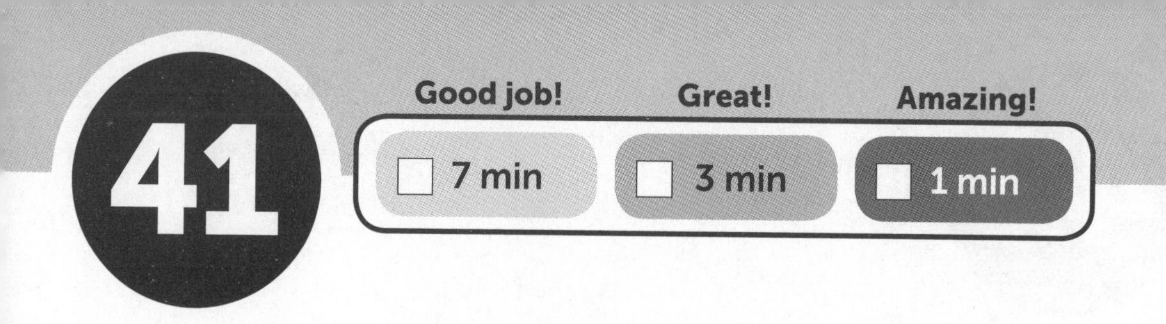
41
Good job!
Great!
Amazing!
7 min
3 min
1 min

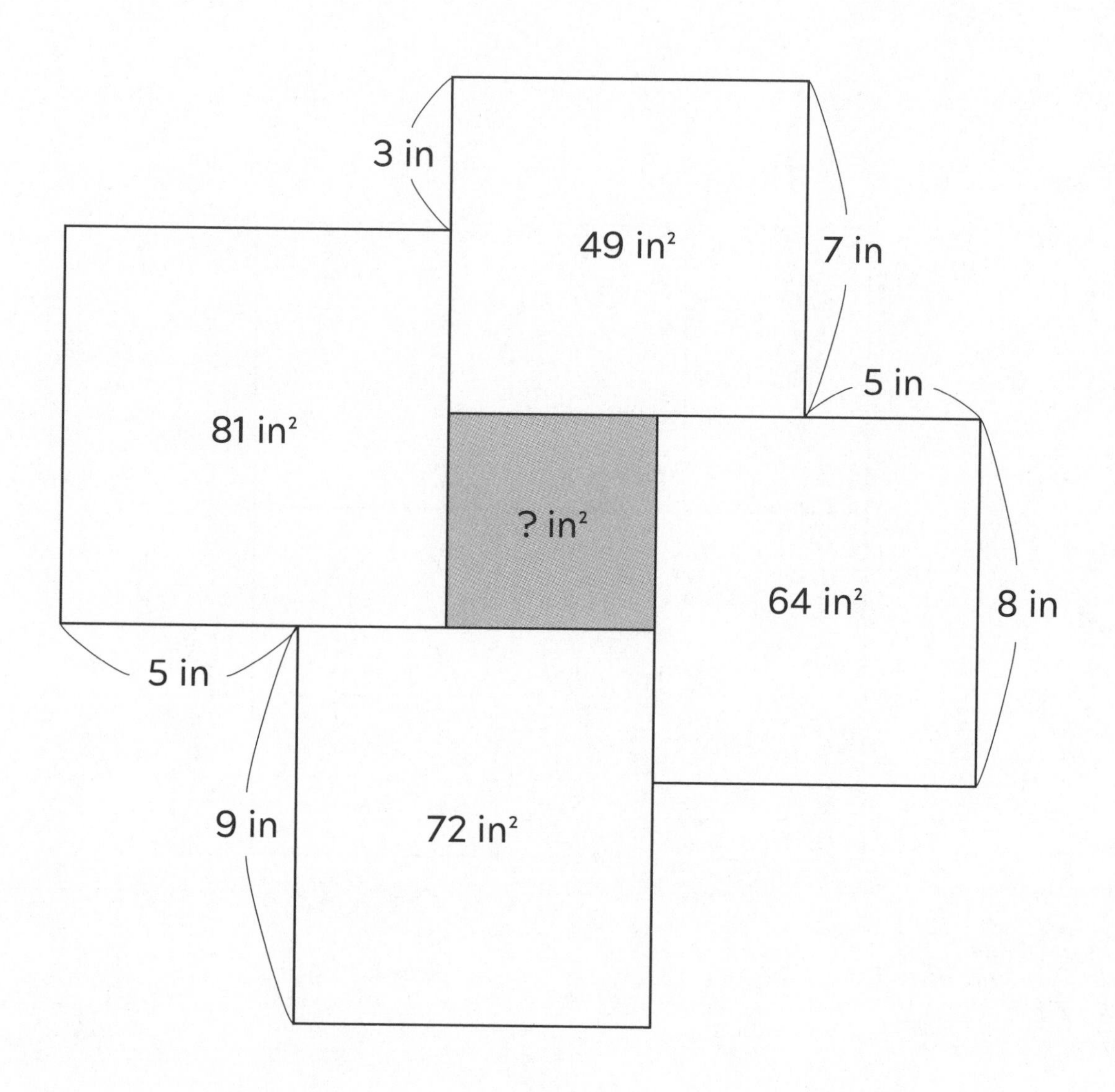
3 in
49 in²
7 in
5 in
81 in²
? in²
64 in²
8 in
5 in
9 in
72 in²

Solution ______________________

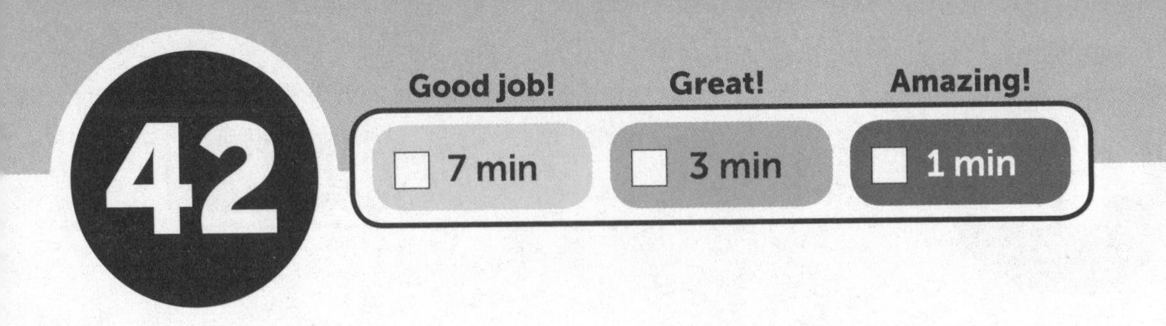
42
Good job!
Great!
Amazing!
7 min
3 min
1 min

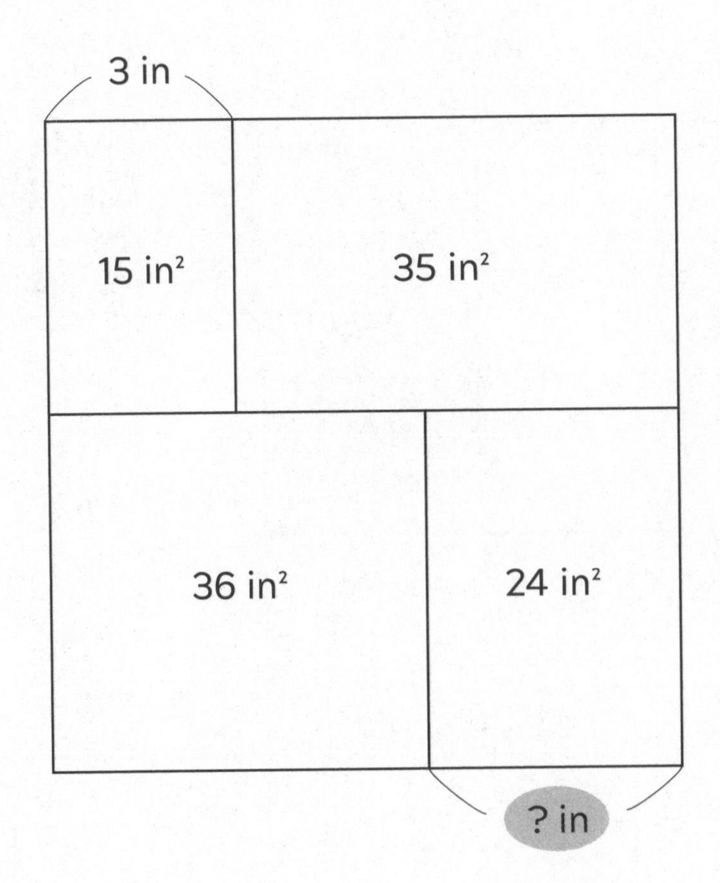
3 in
15 in²
35 in²
36 in²
24 in²
? in

Solution

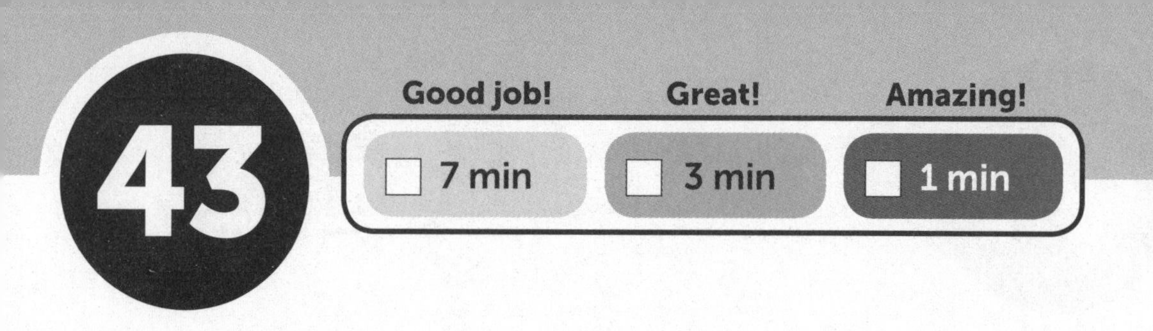
43
Good job!
Great!
Amazing!
7 min
3 min
1 min

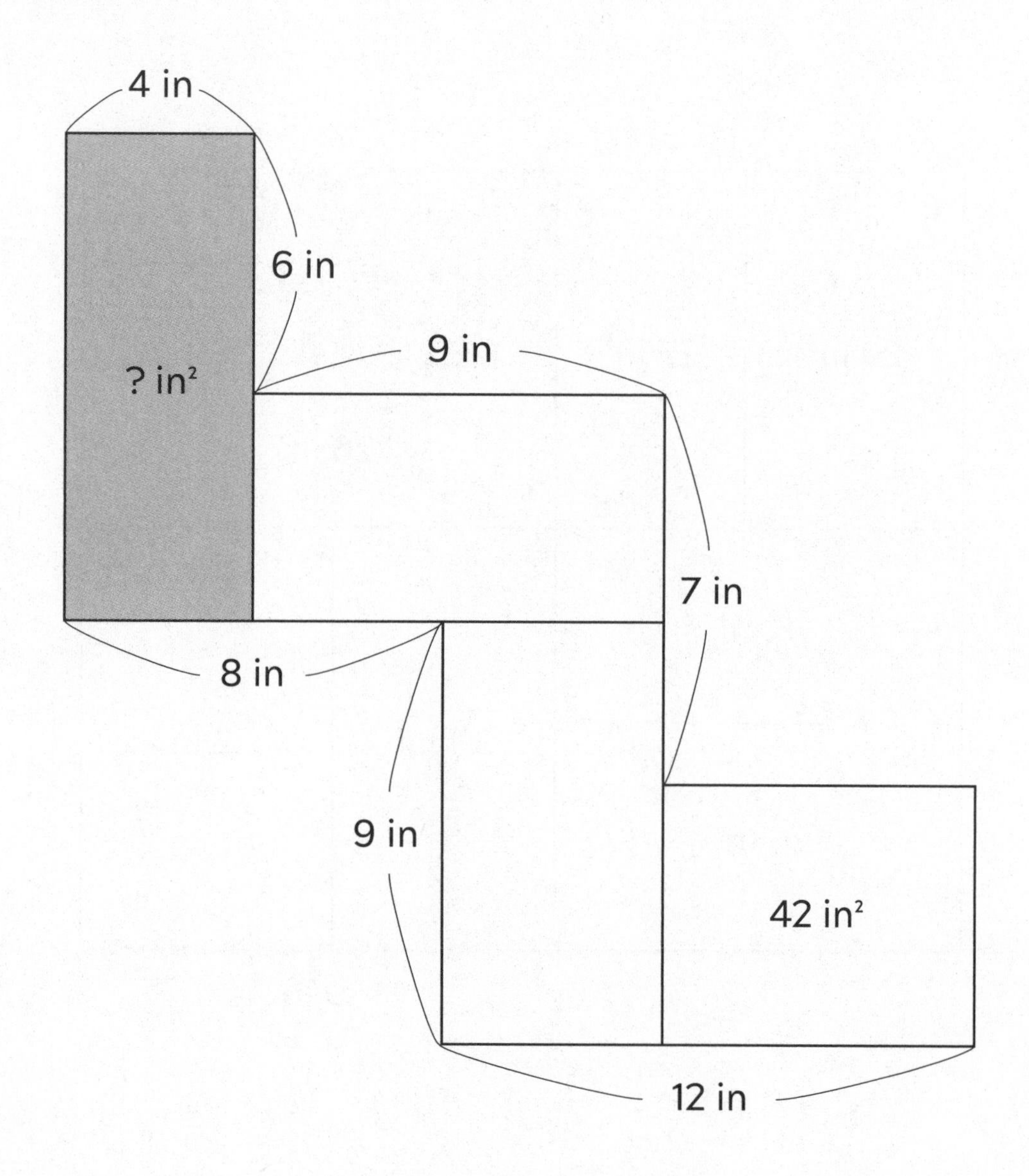
4 in
6 in
9 in
? in²
7 in
8 in
9 in
42 in²
12 in

Solution

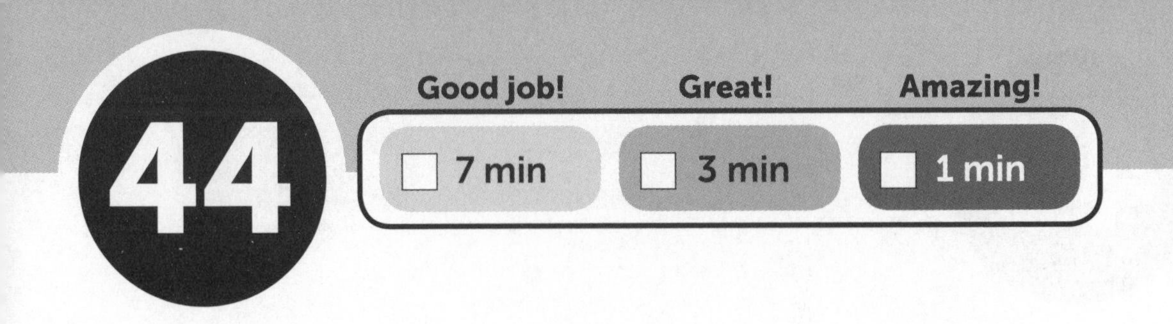

? in²

29 in²

27 in²

26 in²

28 in²

30 in²

42 in²

36 in²

9 in

Solution ________________

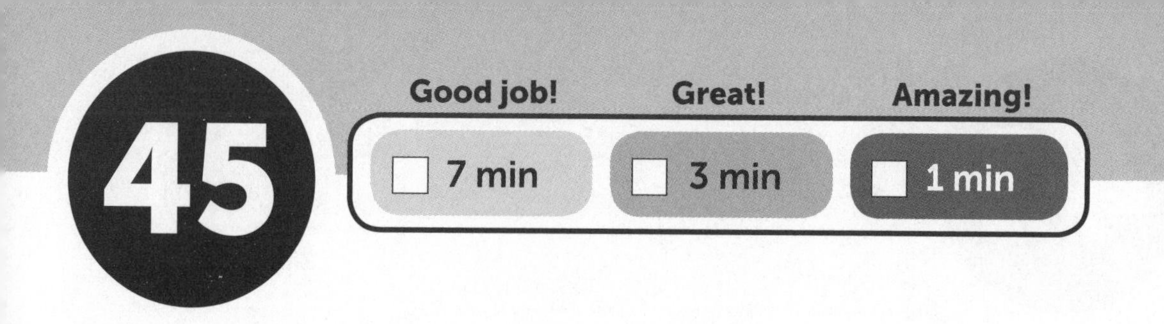
45
Good job!
Great!
Amazing!
7 min
3 min
1 min

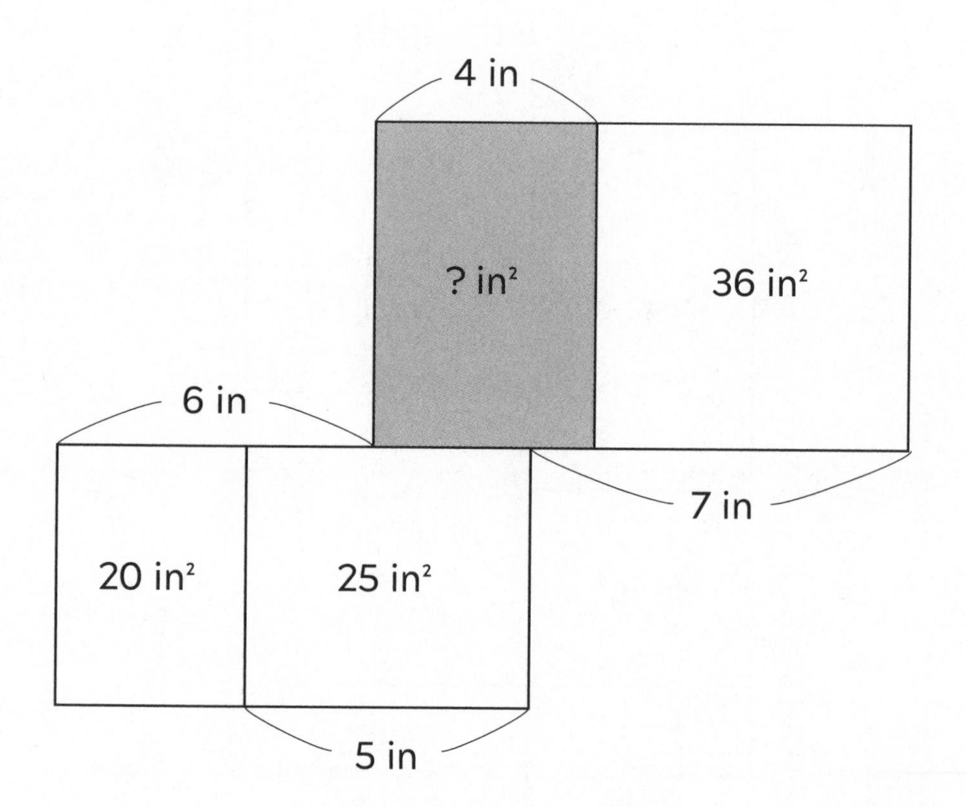
4 in
? in²
36 in²
6 in
7 in
20 in²
25 in²
5 in

Solution

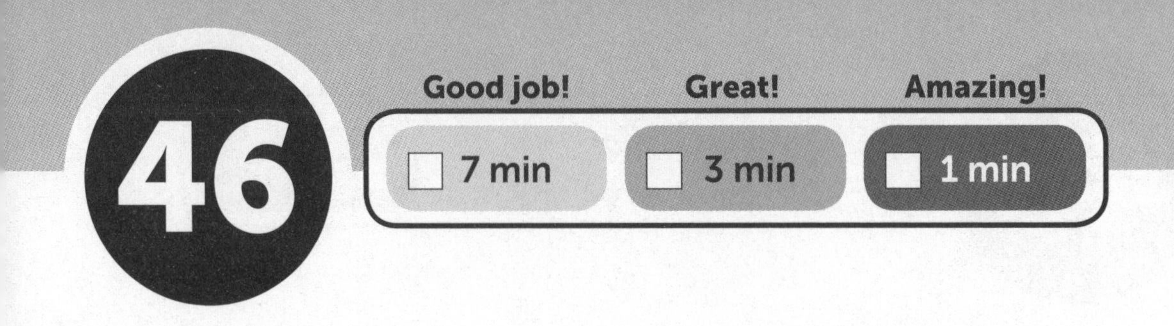
46
Good job!
7 min
Great!
3 min
Amazing!
1 min

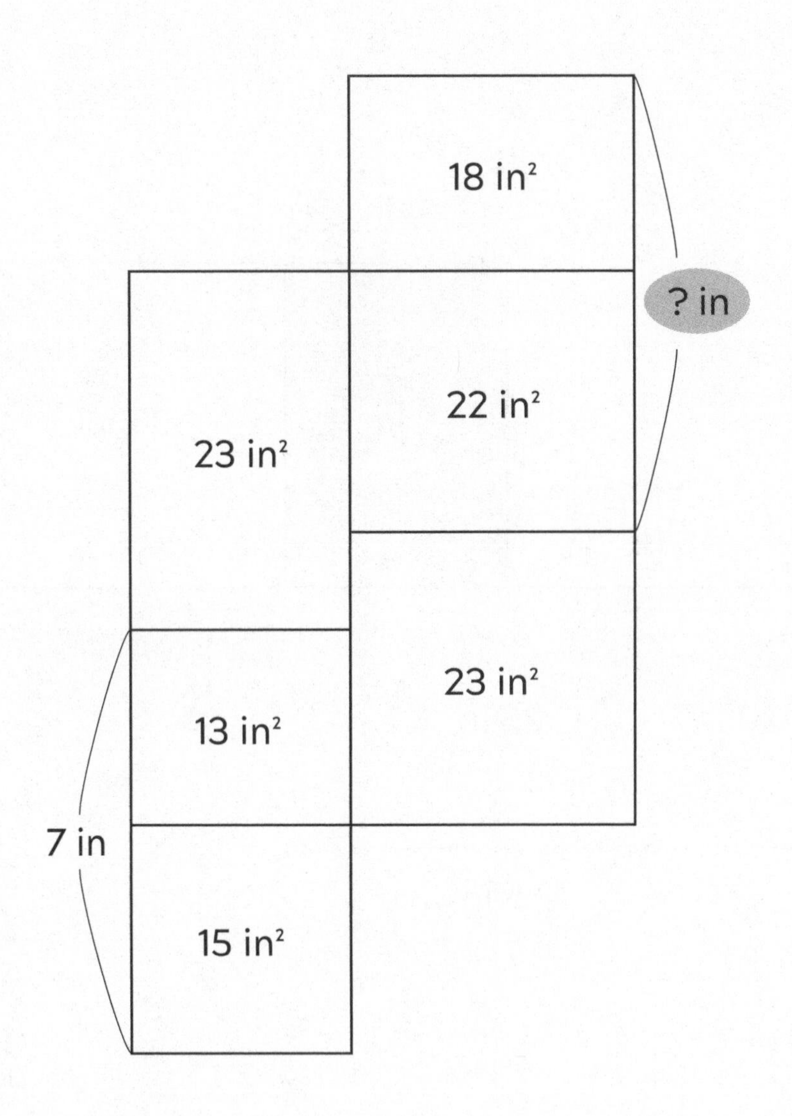
18 in²
? in
22 in²
23 in²
23 in²
13 in²
7 in
15 in²

Solution

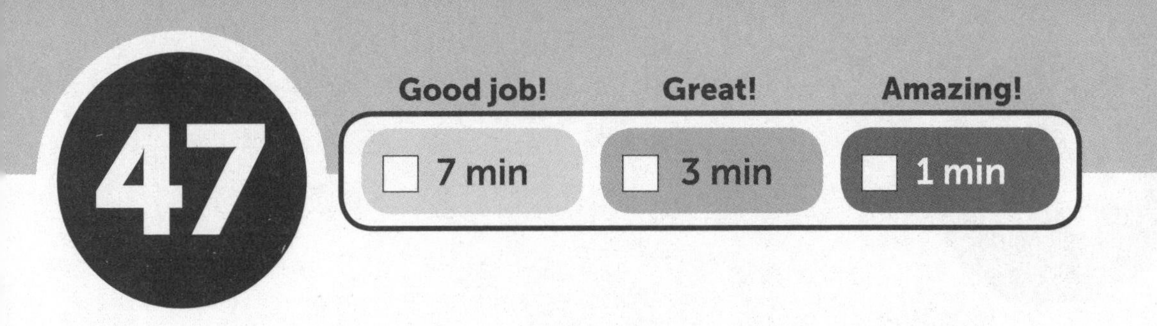
47
Good job!
Great!
Amazing!
7 min
3 min
1 min

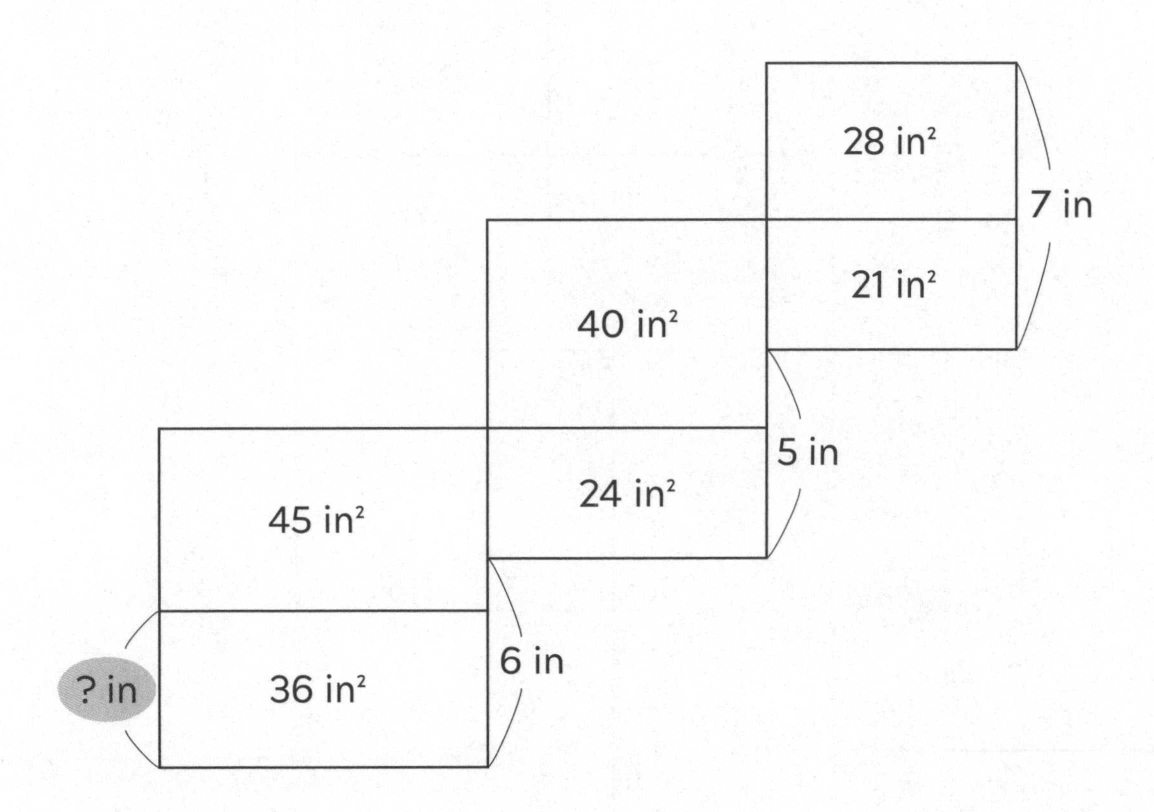
28 in²
7 in
21 in²
40 in²
5 in
24 in²
45 in²
6 in
? in
36 in²

Solution

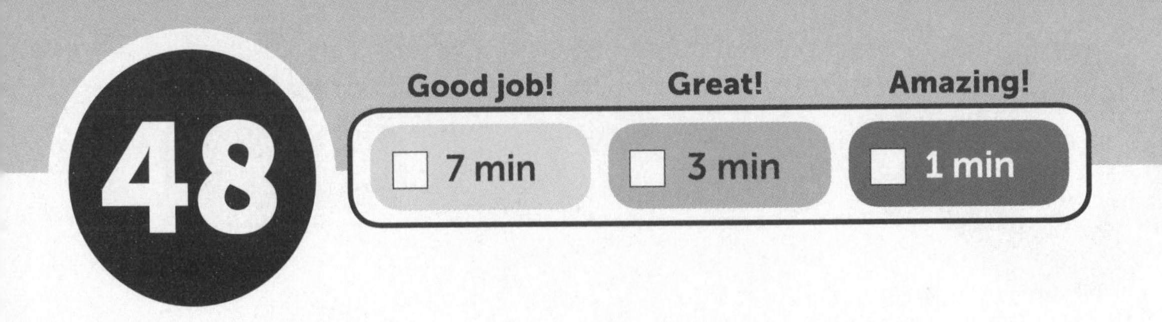
48
Good job!
Great!
Amazing!
7 min
3 min
1 min

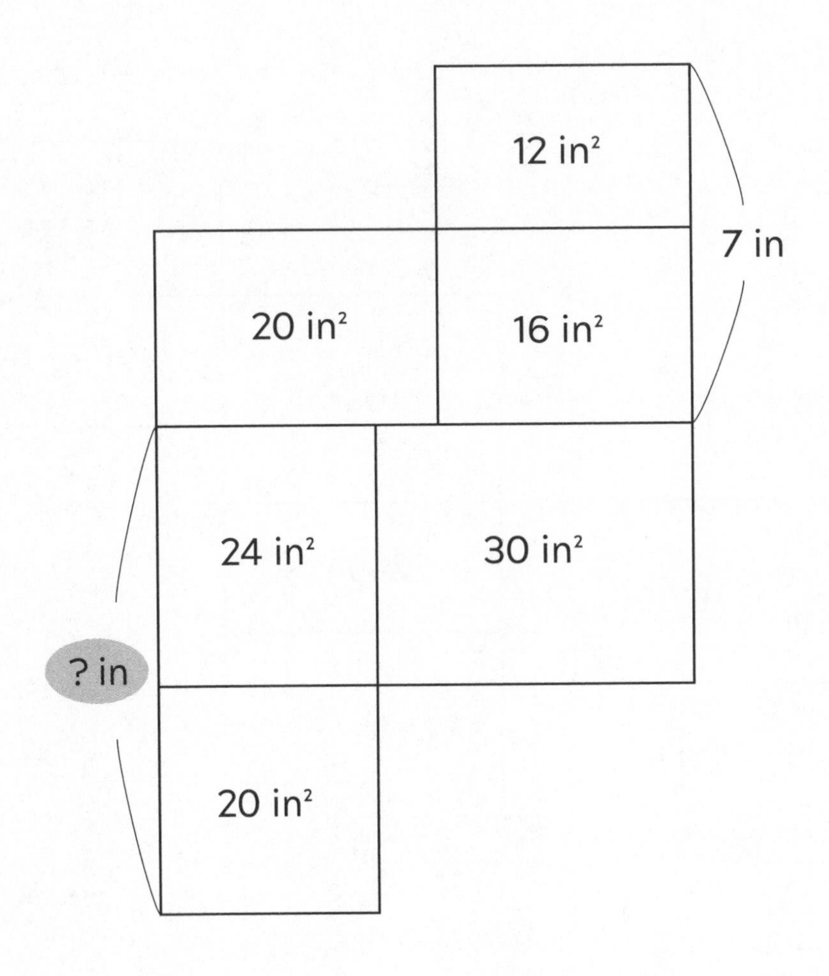
12 in²
7 in
20 in²
16 in²
24 in²
30 in²
? in
20 in²

Solution ______________

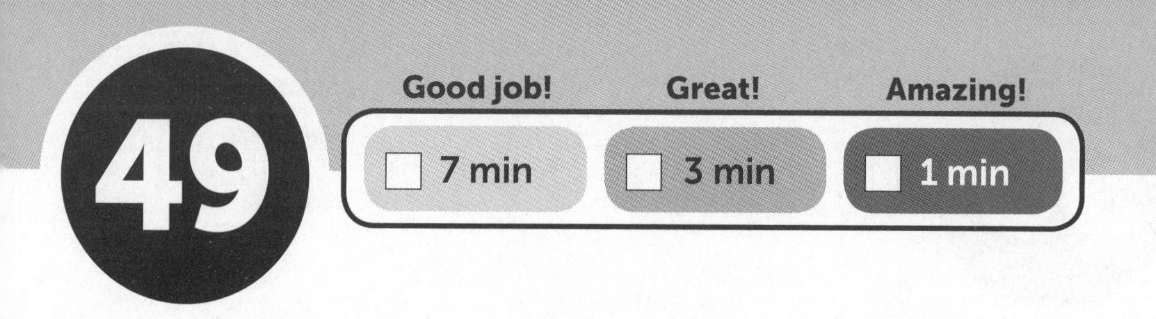
49
Good job!
Great!
Amazing!
7 min
3 min
1 min

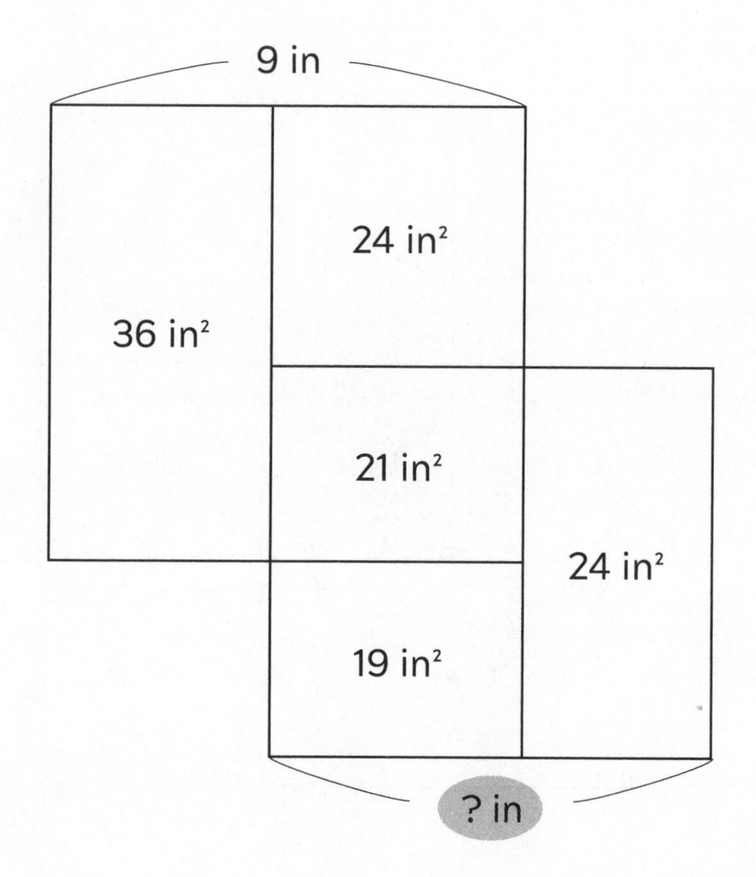
9 in
24 in²
36 in²
21 in²
24 in²
19 in²
? in

Solution ____________________

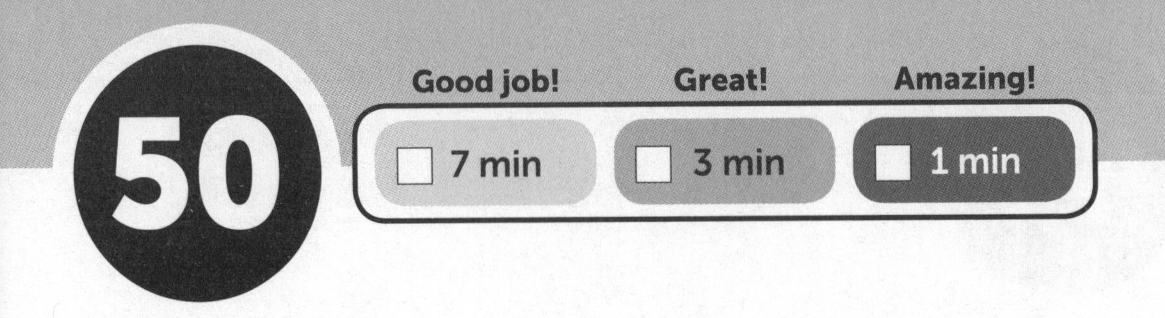
50
Good job!
Great!
Amazing!
7 min
3 min
1 min

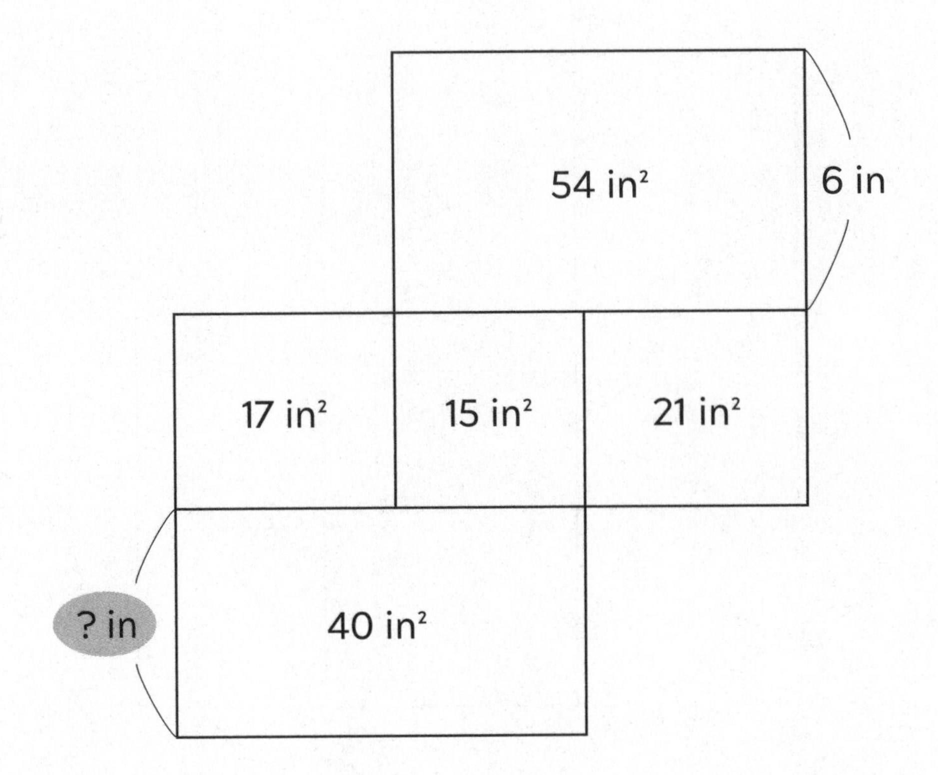
54 in²
6 in
17 in²
15 in²
21 in²
? in
40 in²

Solution

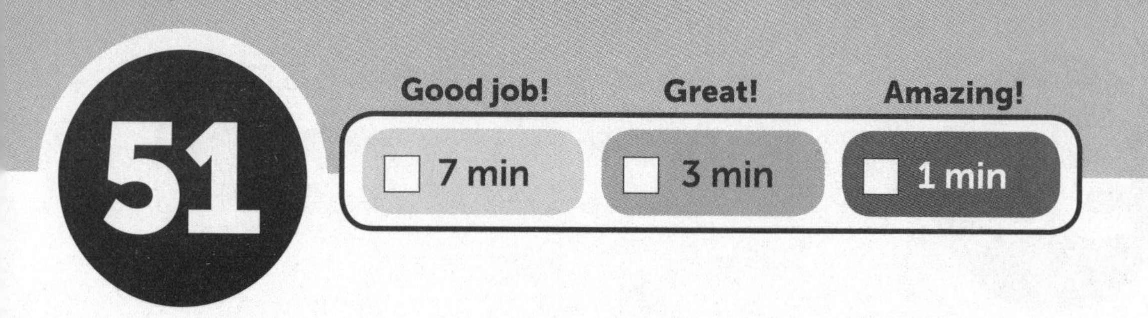
51
Good job!
Great!
Amazing!
7 min
3 min
1 min

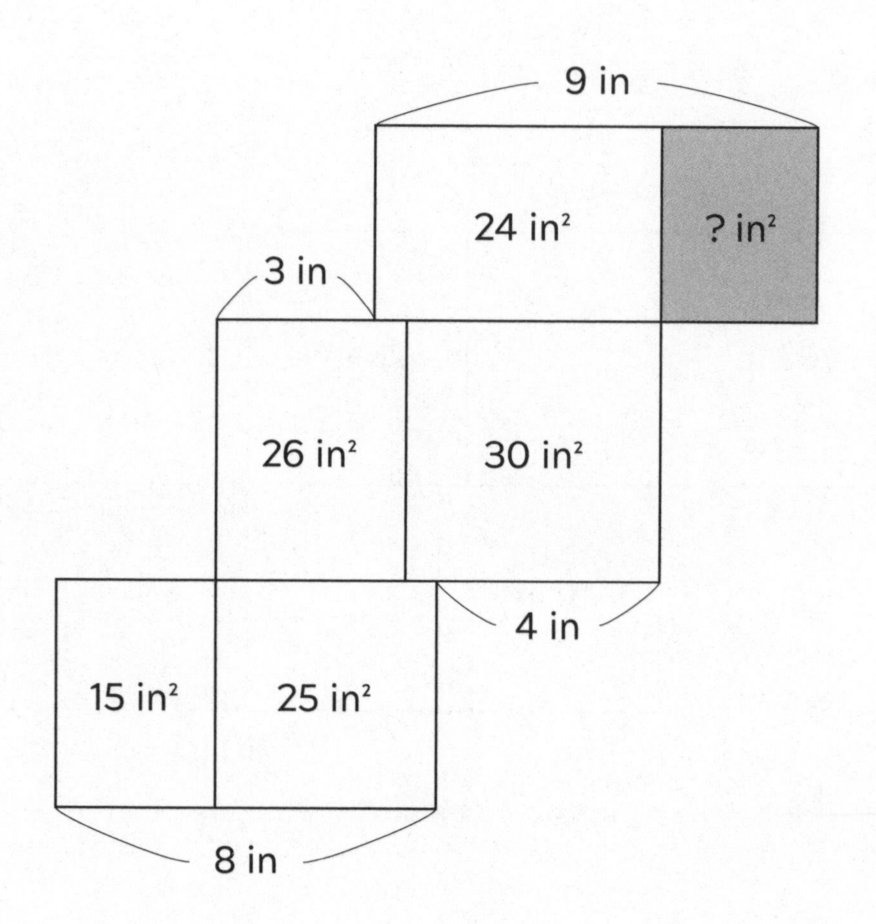
9 in
24 in²
? in²
3 in
26 in²
30 in²
4 in
15 in²
25 in²
8 in

Solution

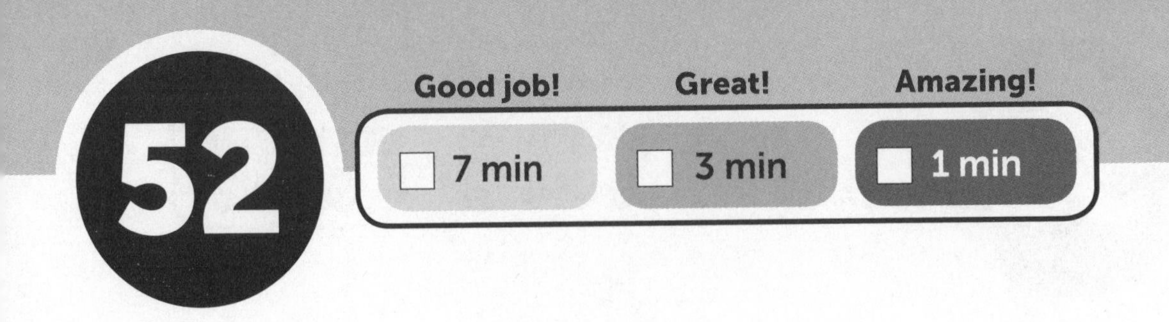
52
Good job!
Great!
Amazing!
7 min
3 min
1 min

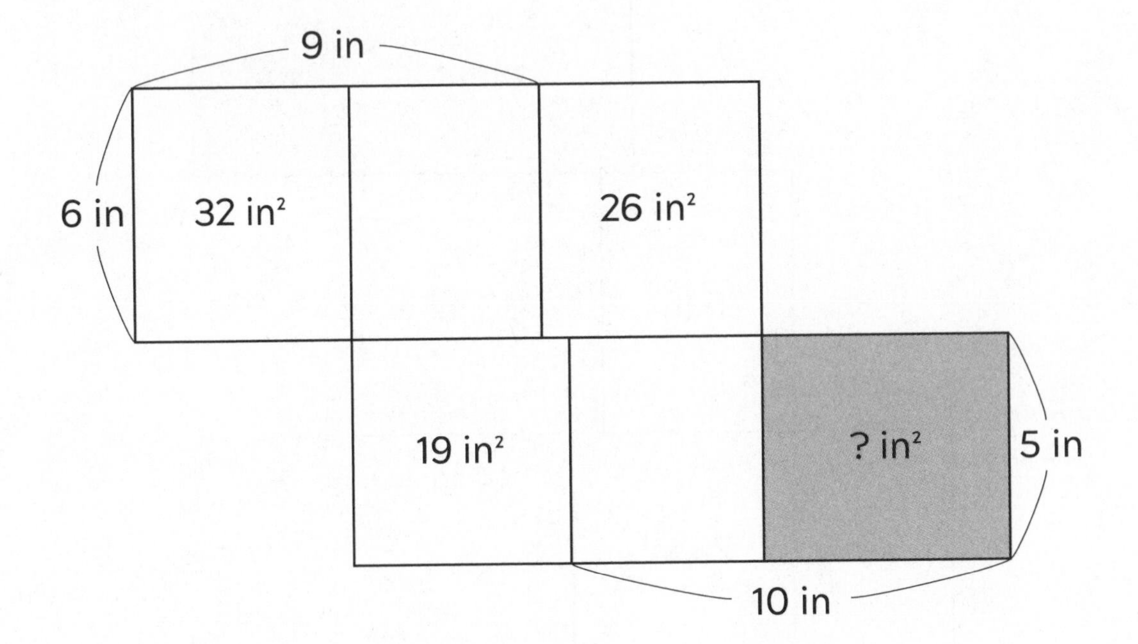
9 in
6 in
32 in²
26 in²
19 in²
? in²
5 in
10 in

Solution ____________________

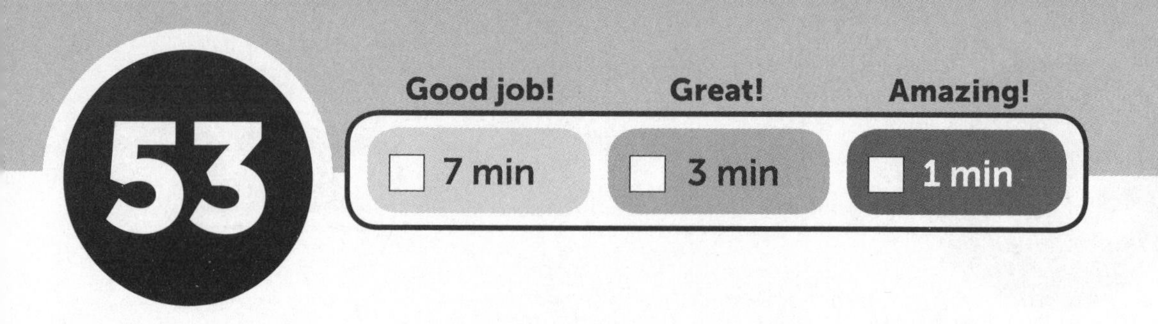
53
Good job!
Great!
Amazing!
7 min
3 min
1 min

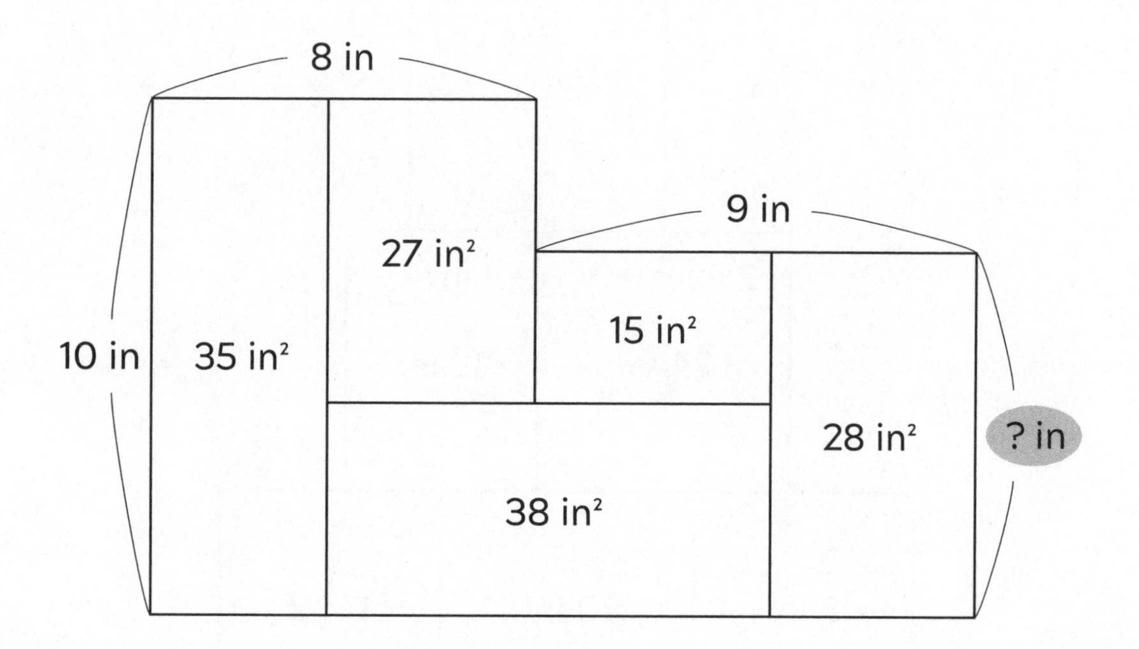
8 in
9 in
27 in²
15 in²
10 in
35 in²
28 in²
? in
38 in²

Solution ____________________

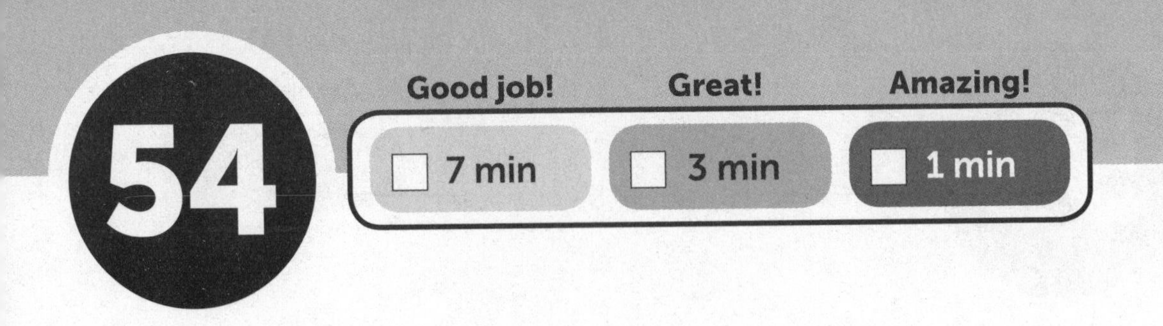
54
Good job!
Great!
Amazing!
7 min
3 min
1 min

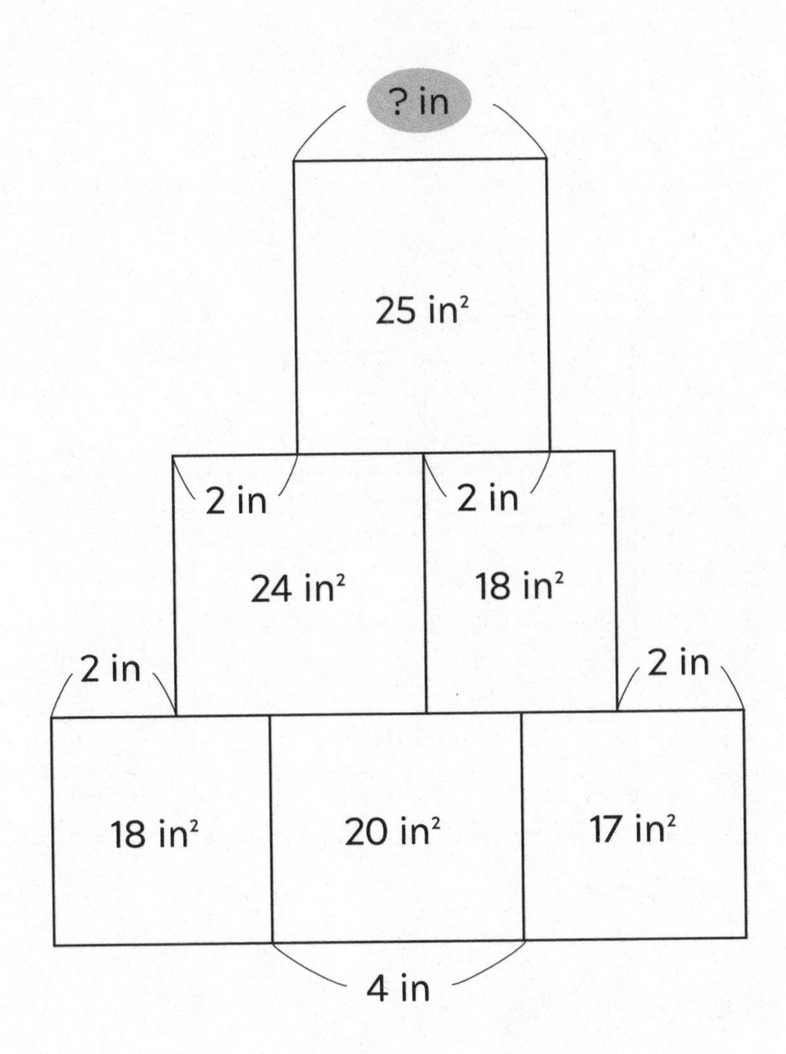
? in
25 in²
2 in
2 in
24 in²
18 in²
2 in
2 in
18 in²
20 in²
17 in²
4 in

Solution ____________________

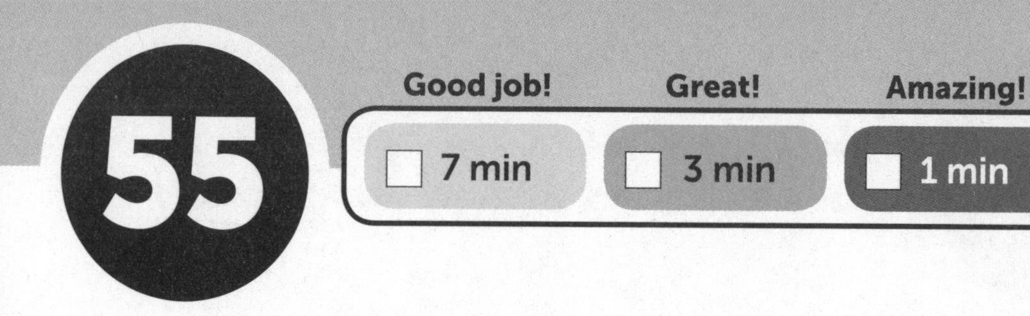

2 in

4 in²

6 in²

3 in²

9 in²

20 in²

4 in²

8 in²

10 in²

7 in²

6 in²

15 in²

6 in²

12 in²

? in

14 in²

11 in²

8 in²

16 in²

Solution ________________

56

Good job! ☐ 7 min

Great! ☐ 3 min

Amazing! ☐ 1 min

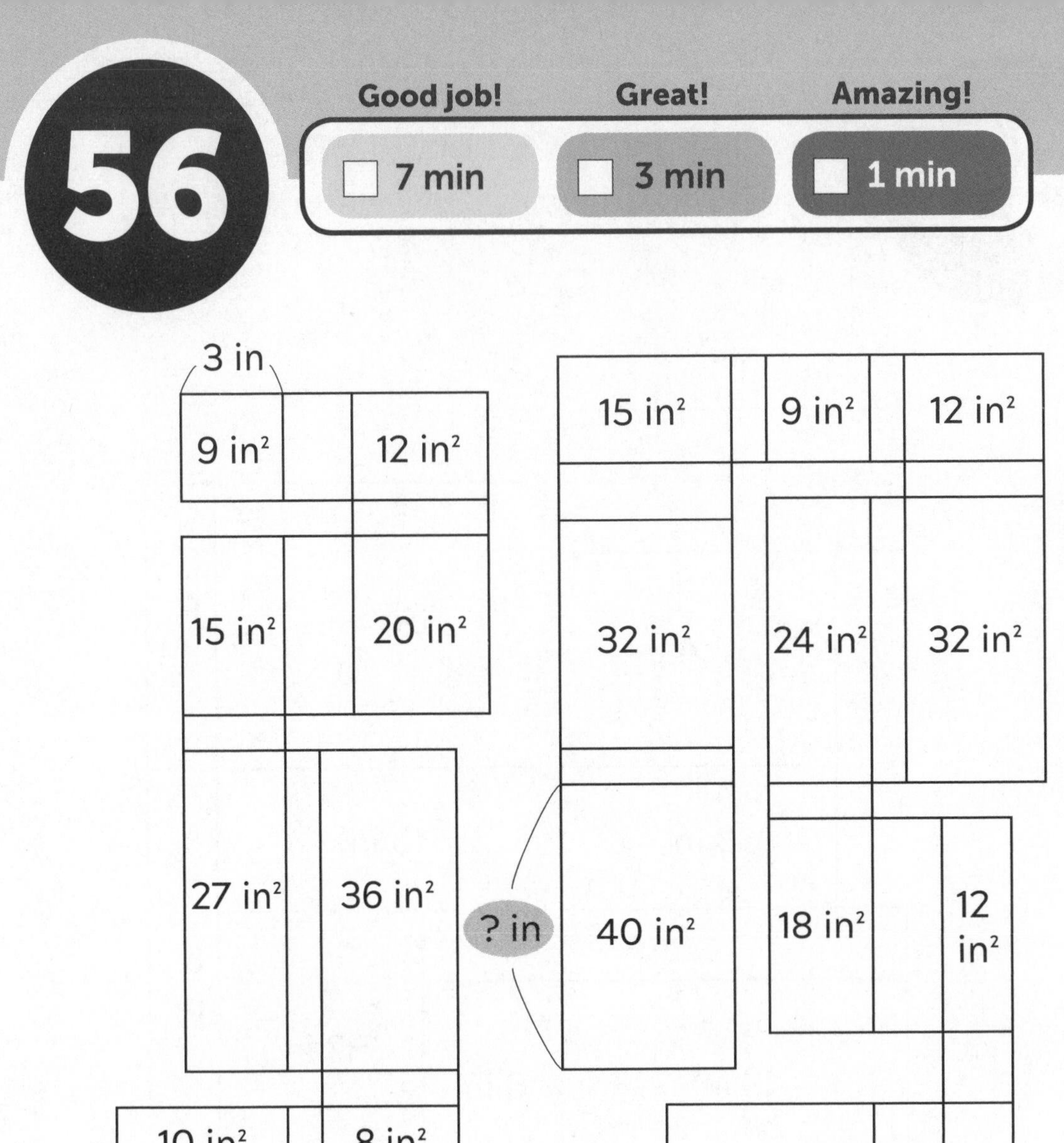

Solution ____________

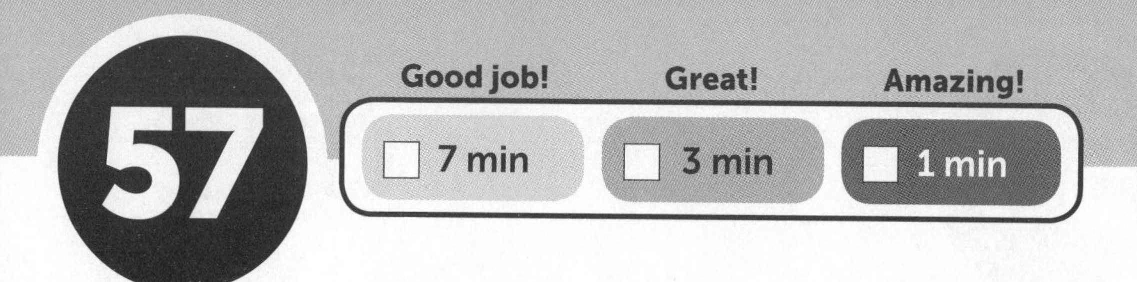

? in

8 in

50 in²

42 in²

56 in²

32 in²

60 in²

54 in²

63 in²

20 in²

35 in²

49 in²

30 in²

60 in²

24 in²

25 in²

40 in²

36 in²

66 in²

48 in²

Solution ____________

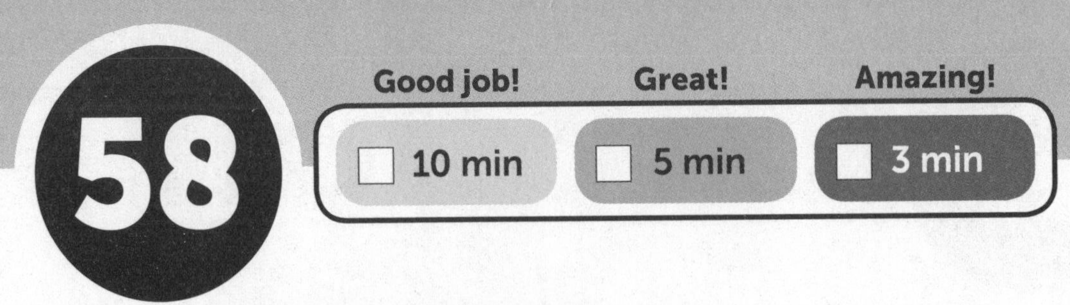

? in^2

25 in^2

35 in^2

56 in^2

48 in^2

42 in^2

16 in^2

12 in^2

49 in^2

21 in^2

18 in^2

28 in^2

54 in^2

72 in^2

35 in^2

30 in^2

24 in^2

40 in^2

45 in^2

30 in^2

9 in

Solution ____________

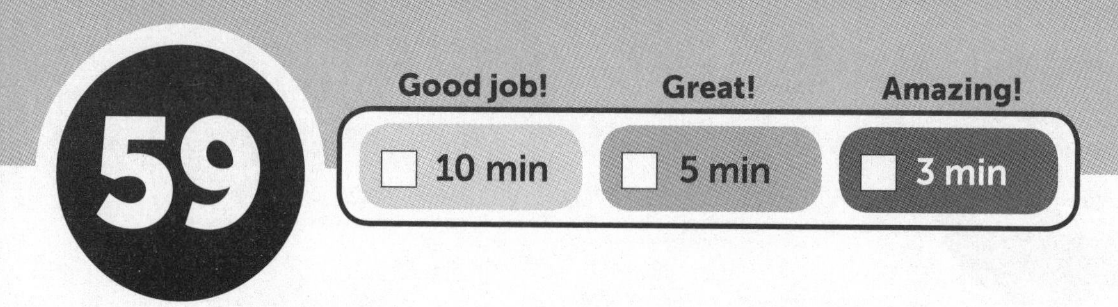
59
Good job!
Great!
Amazing!
10 min
5 min
3 min

? in²
15 in²
36 in²
15 in²
12 in²
63 in²
28 in²
42 in²
14 in²
12 in²
16 in²
49 in²
56 in²
27 in²
45 in²
35 in²
25 in²
24 in²
20 in²
15 in²
40 in²
18 in²
21 in²
24 in²
7 in

Solution ______________________

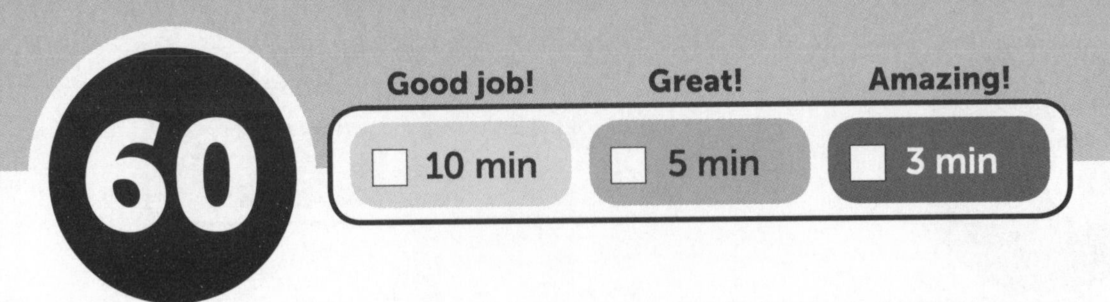

20 in^2 | 14 in^2 | 24 in^2 | 20 in^2 | 40 in^2

25 in^2 | 20 in^2 | 18 in^2 | 12 in^2 | 20 in^2

21 in^2 | 12 in^2 | ? in^2 | 18 in^2 | 48 in^2

20 in^2

36 in^2 | 18 in^2 | 30 in^2 | 21 in^2 | 49 in^2

54 in^2 | 36 in^2 | 40 in^2 | 32 in^2 | 42 in^2

20 in^2 | 16 in^2 | 21 in^2 | 20 in^2 | 35 in^2

5 in

Solution __________________

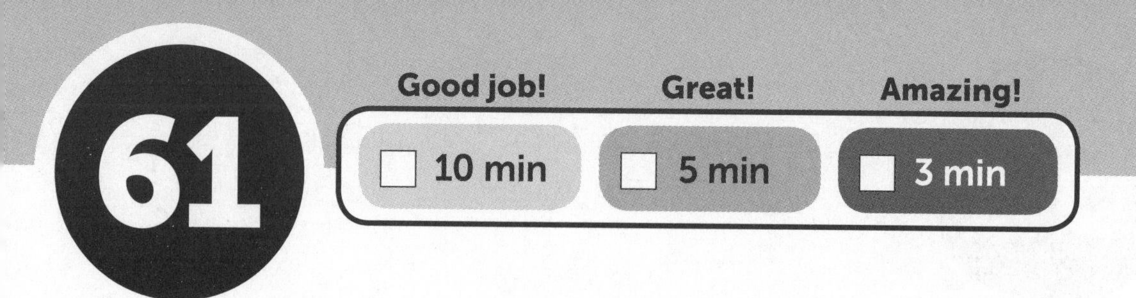
61
Good job!
Great!
Amazing!
10 min
5 min
3 min

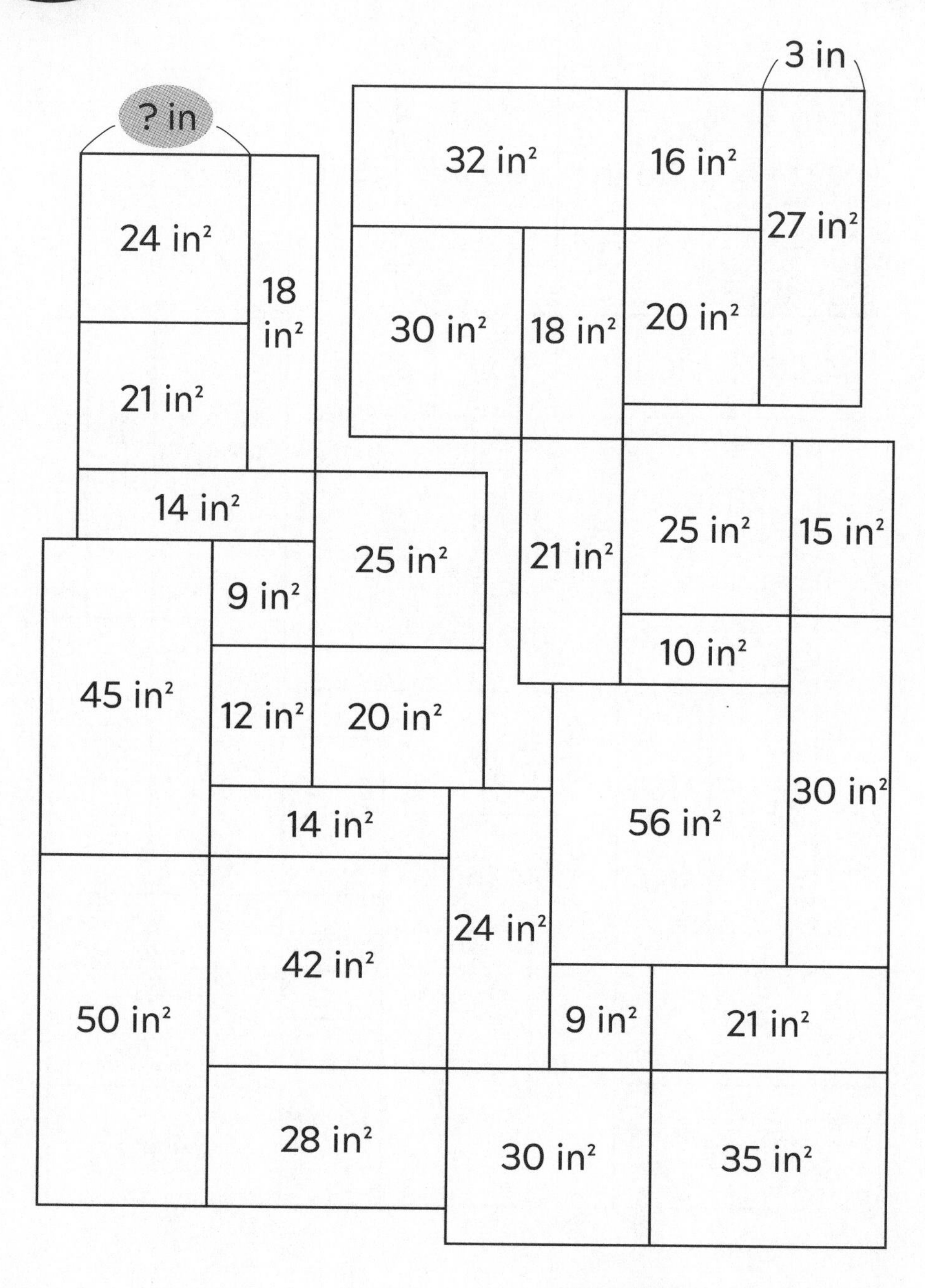
3 in
? in
32 in²
16 in²
27 in²
24 in²
18 in²
30 in²
18 in²
20 in²
21 in²
14 in²
25 in²
21 in²
25 in²
15 in²
9 in²
10 in²
45 in²
12 in²
20 in²
14 in²
56 in²
30 in²
24 in²
42 in²
9 in²
21 in²
50 in²
28 in²
30 in²
35 in²

Solution

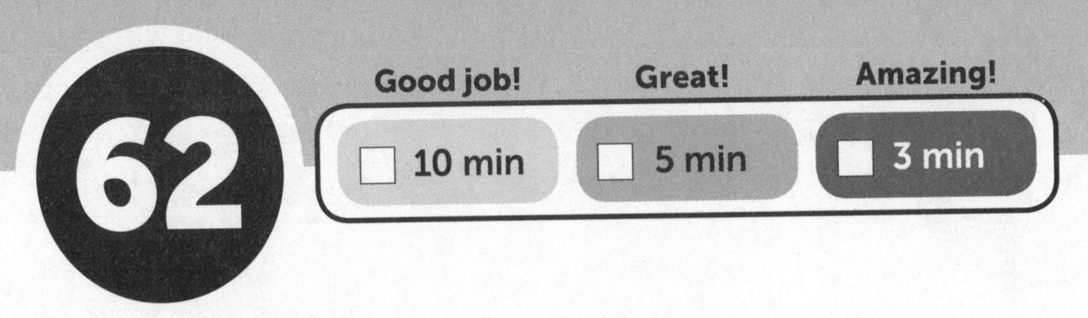

? in^2
36 in^2
39 in^2
25 in^2
27 in^2
21 in^2
14 in^2
16 in^2
11 in^2
13 in^2
36 in^2
18 in^2
28 in^2
35 in^2
45 in^2
20 in^2
24 in^2
18 in^2
25 in^2
24 in^2
14 in^2
32 in^2
24 in^2
36 in^2
21 in^2
35 in^2
11 in^2
39 in^2
28 in^2
16 in^2
24 in^2
8 in

Solution ________________

challenge puzzles

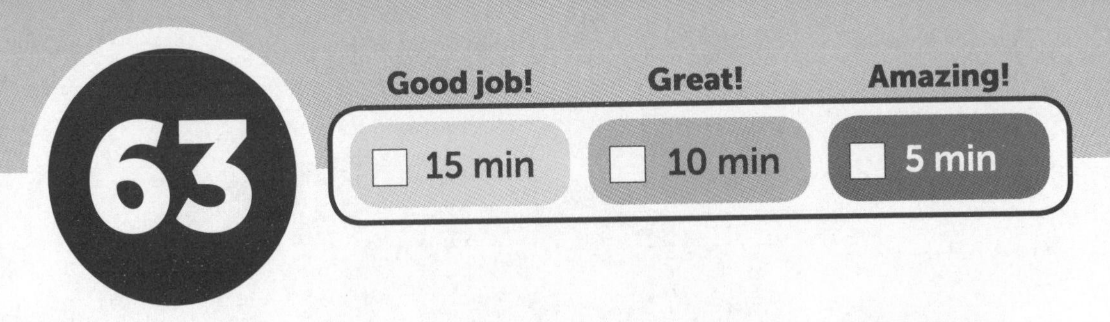

27 in²
30 in²
? in²
9 in²
20 in²
14 in²
54 in²
40 in²
36 in²
12 in²
14 in²
42 in²
30 in²
32 in²
28 in²
20 in²
18 in²
28 in²
14 in²
25 in²
35 in²
21 in²
16 in²
24 in²
30 in²
25 in²
42 in²
36 in²
6 in

Solution

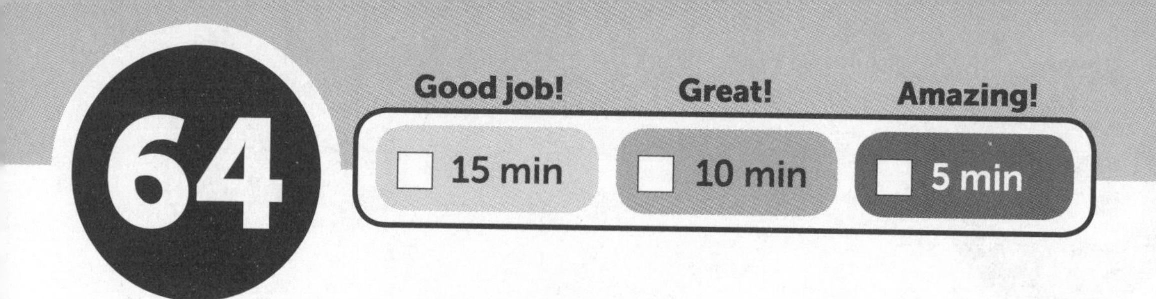

15 in² ? in² 16 in² 72 in²

30 in² 48 in² 20 in²

21 in² 12 in² 40 in² 10 in²

16 in² 24 in² 25 in² 5 in 81 in² 18 in²

20 in² 36 in² 30 in²

15 in²

100 in² 35 in² 32 in² 30 in²

27 in² 18 in²

Solution ____________

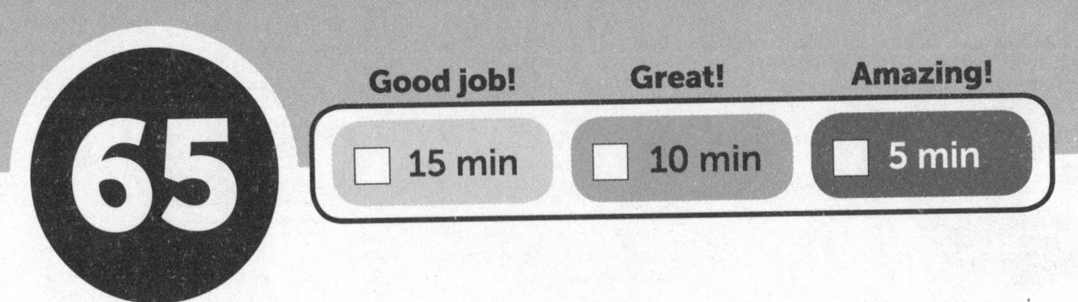

28 in^2
40 in^2
? in
60 in^2
27 in^2
9 in^2
12 in^2
16 in^2
10 in^2
24 in^2
15 in^2
24 in^2
32 in^2
24 in^2
18 in^2
20 in^2
42 in^2
16 in^2
35 in^2
12 in^2
28 in^2
21 in^2
49 in^2
25 in^2
36 in^2
30 in^2
12 in^2
4 in
28 in^2
24 in^2
16 in^2
20 in^2

Solution ____________

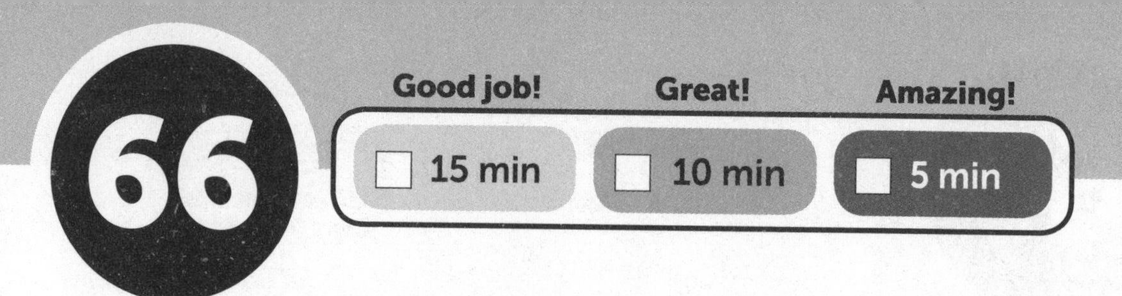

3 in

9 in²

36 in²

25 in²

24 in²

7 in²

10 in²

21 in²

15 in²

12 in²

20 in²

30 in²

9 in²

15 in²

28 in²

10 in²

24 in²

42 in²

18 in²

24 in²

20 in²

6 in²

24 in²

20 in²

26 in²

44 in²

22 in²

27 in²

? in²

12 in²

12 in²

7 in²

6 in²

Solution ________________

67

Good job! — 15 min
Great! — 10 min
Amazing! — 5 min

49 in²
56 in²
24 in²
12 in²
18 in²
36 in²
81 in²
28 in²
20 in²
24 in²
32 in²
15 in²
42 in²
35 in²
40 in²
30 in²
42 in²
12 in²
5 in
30 in²
14 in²
? in²
16 in²
36 in²
21 in²
20 in²
45 in²
20 in²
20 in²
42 in²
25 in²
12 in²
24 in²
10 in²
35 in²
24 in²
18 in²
42 in²
30 in²
64 in²
72 in²
40 in²

Solution ____________

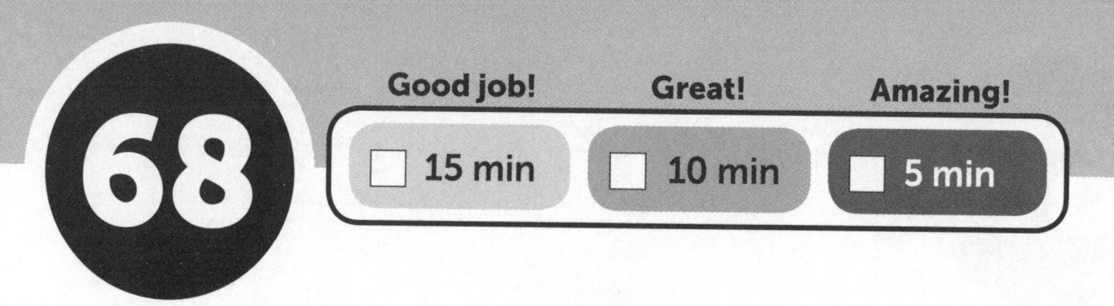

? in

60 in² 63 in² 18 in² 20 in² 72 in²

48 in² 15 in² 12 in² 14 in² 81 in²

64 in² 14 in² 28 in² 16 in² 49 in²

14 in² 16 in²

45 in² 56 in² 10 in² 27 in² 40 in² 42 in²

21 in² 36 in² 32 in²

48 in² 18 in² 40 in² 16 in² 12 in² 25 in²

30 in²

24 in² 25 in² 30 in² 20 in² 35 in² 21 in² 36 in²

15 in² 24 in² 28 in² 20 in² 9 in² 18 in²

7 in

Solution ____________________

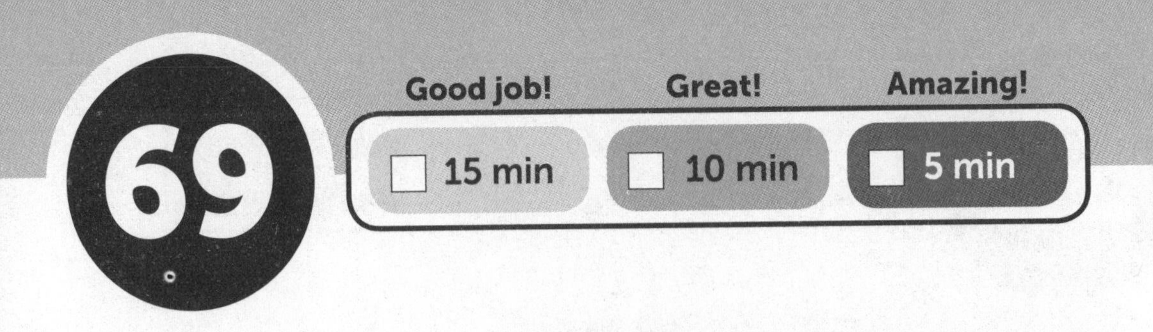

56 in²
? in²
27 in²
63 in²
18 in²
42 in²
90 in²
12 in²
40 in²
66 in²
60 in²
36 in²
9 in²
27 in²
28 in²
32 in²
4 in
16 in²
21 in²
30 in²
16 in²
10 in²
12 in²
52 in²
80 in²
24 in²
7 in²
30 in²
70 in²
33 in²
14 in²
45 in²
24 in²
64 in²
24 in²
32 in²
10 in²
15 in²
48 in²
12 in²
14 in²

Solution ____________

70

Good job! Great! Amazing!

☐ 15 min ☐ 10 min ☐ 5 min

7 in

35 in², 25 in², 20 in², 16 in², 24 in², 27 in², 56 in², 30 in², 21 in², 18 in², 20 in², 18 in², 15 in², 28 in², 45 in², 42 in², 36 in², 32 in², 30 in², 30 in², 12 in², 42 in², 36 in², 10 in², 24 in², 36 in², 12 in², 27 in², 21 in², 49 in², 27 in², 21 in², 24 in², 10 in², 10 in², 40 in², 15 in², 45 in², ? in², 48 in², 18 in², 28 in², 35 in², 20 in²

Solution ________________

solutions

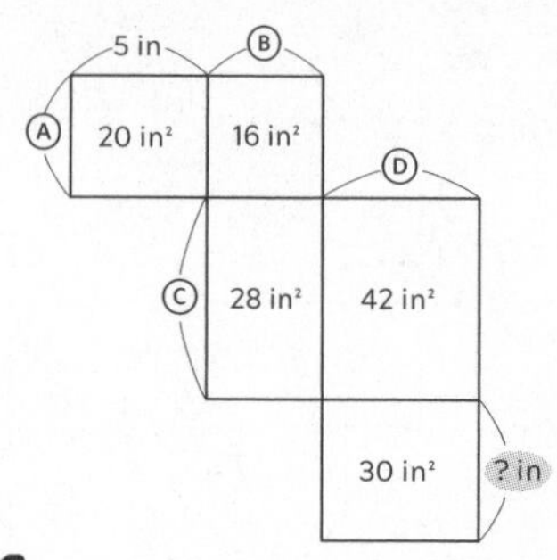

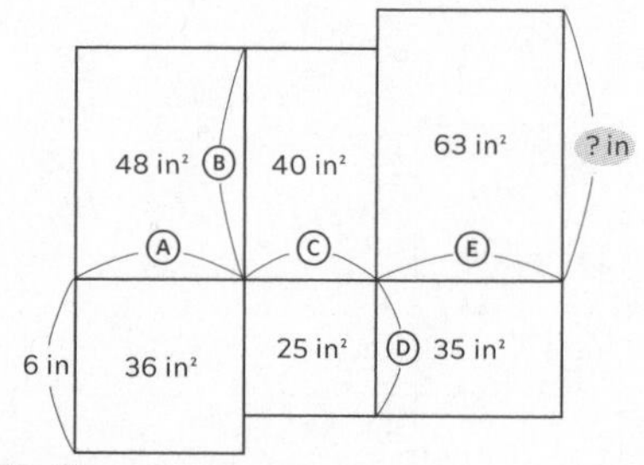

Puzzle 1

Ⓐ . . . 20 ÷ 5 = 4 in.
Ⓑ . . . 16 ÷ Ⓐ = 4 in.
Ⓒ . . . 28 ÷ Ⓑ = 7 in.
Ⓓ . . . 42 ÷ Ⓒ = 6 in.
Length ? is 30 ÷ Ⓓ = 5 in.

Puzzle 2

Ⓐ . . . 36 ÷ 6 = 6 in.
Ⓑ . . . 48 ÷ Ⓐ = 8 in.
Ⓒ . . . 40 ÷ Ⓑ = 5 in.
Ⓓ . . . 25 ÷ Ⓒ = 5 in.
Ⓔ . . . 35 ÷ Ⓓ = 7 in.
Length ? is 63 ÷ Ⓔ = 9 in.

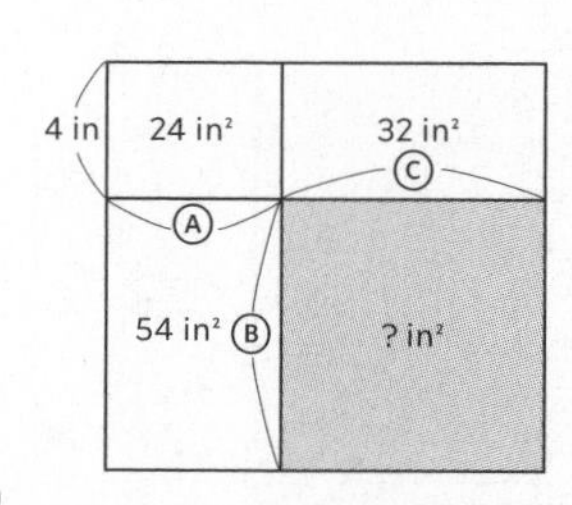

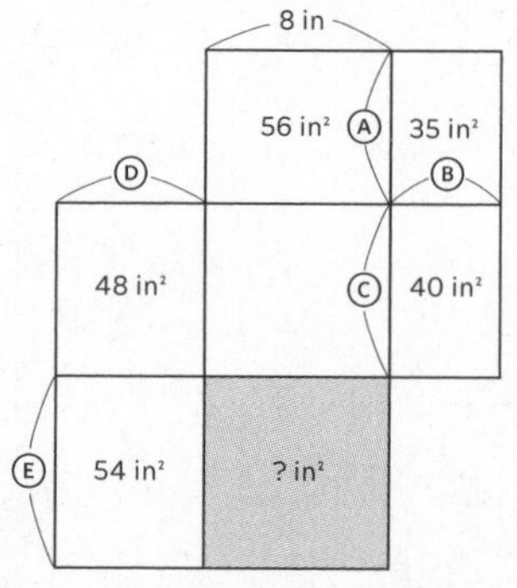

Puzzle 3

Ⓐ . . . 24 ÷ 4 = 6 in.
Ⓑ . . . 54 ÷ Ⓐ = 9 in.
Ⓒ . . . 32 ÷ 4 = 8 in.
Area ? is Ⓑ × Ⓒ = 72 in.²

Puzzle 4

Ⓐ . . . 56 ÷ 8 = 7 in.
Ⓑ . . . 35 ÷ Ⓐ = 5 in.
Ⓒ . . . 40 ÷ Ⓑ = 8 in.
Ⓓ . . . 48 ÷ Ⓒ = 6 in.
Ⓔ . . . 54 ÷ Ⓓ = 9 in.
Area ? is 8 × Ⓔ = 72 in.²

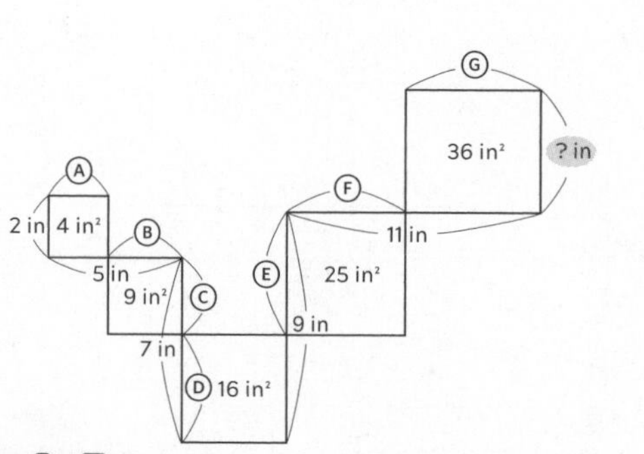

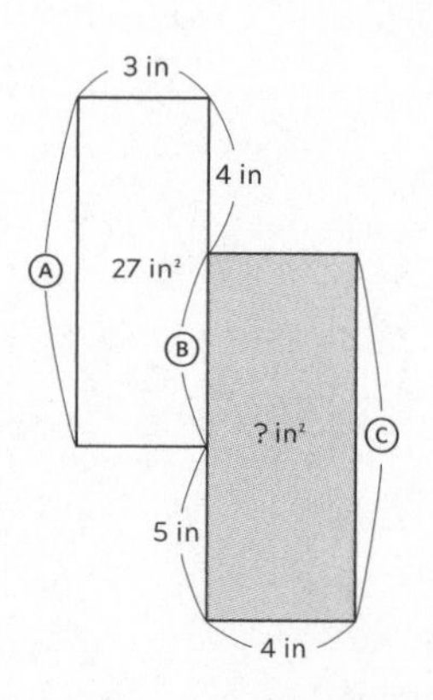

Puzzle 5

Ⓐ . . . 4 ÷ 2 = 2 in.
Ⓑ . . . 5 – Ⓐ = 3 in.
Ⓒ . . . 9 ÷ Ⓑ = 3 in.
Ⓓ . . . 7 – Ⓒ = 4 in.
Ⓔ . . . 9 – Ⓓ = 5 in.
Ⓕ . . . 25 ÷ Ⓔ = 5 in.
Ⓖ . . . 11 – Ⓕ = 6 in.
Length ? is 36 ÷ Ⓖ = 6 in.

Puzzle 6

Ⓐ . . . 27 ÷ 3 = 9 in.
Ⓑ . . . Ⓐ – 4 = 5 in.
Ⓒ . . . Ⓑ + 5 = 10 in.
Area ? is Ⓒ × 4 = 40 in.²

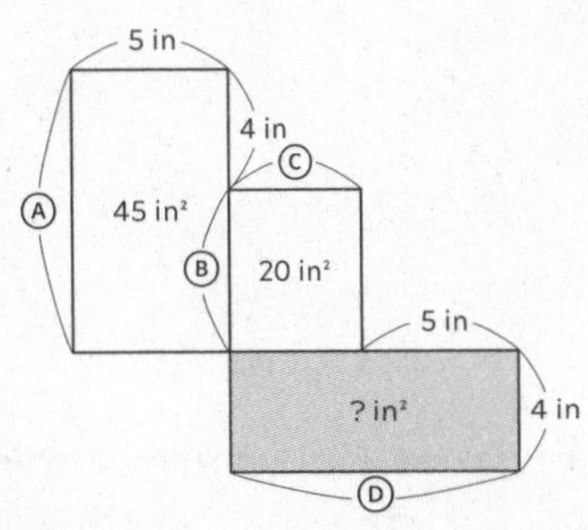

Puzzle 7

Ⓐ . . . 45 ÷ 5 = 9 in.
Ⓑ . . . Ⓐ – 4 = 5 in.
Ⓒ . . . 20 ÷ Ⓑ = 4 in.
Ⓓ . . . Ⓒ + 5 = 9 in.
Area ? is Ⓓ × 4 = 36 in.²

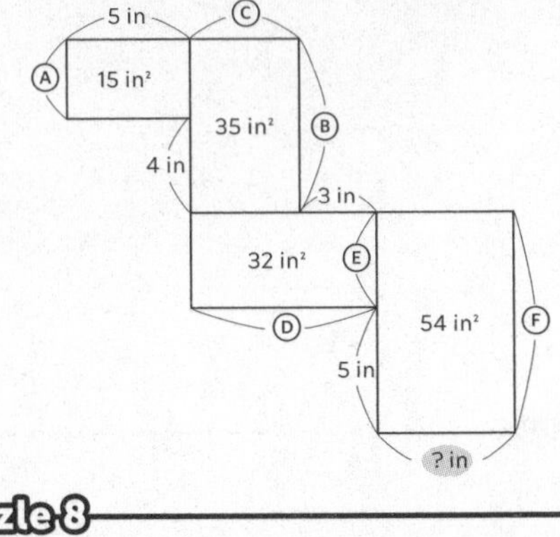

Puzzle 8

Ⓐ . . . 15 ÷ 5 = 3 in.
Ⓑ . . . Ⓐ + 4 = 7 in.
Ⓒ . . . 35 ÷ Ⓑ = 5 in.
Ⓓ . . . Ⓒ + 3 = 8 in.
Ⓔ . . . 32 ÷ Ⓓ = 4 in.
Ⓕ . . . Ⓔ + 5 = 9 in.
Length ? is 54 ÷ Ⓕ = 6 in.

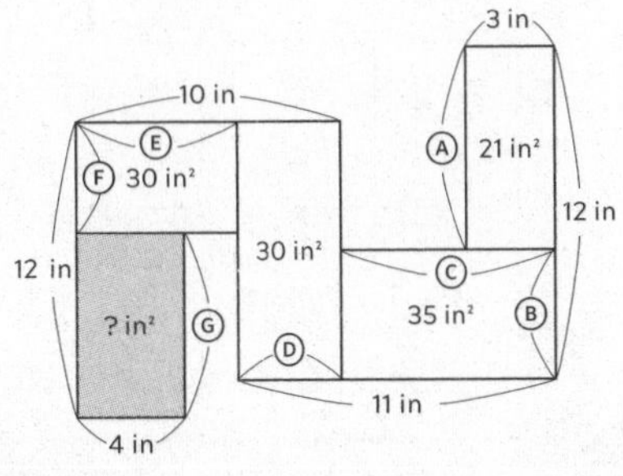

Puzzle 9

Ⓐ . . . 21 ÷ 3 = 7 in.
Ⓑ . . . 12 – Ⓐ = 5 in.
Ⓒ . . . 35 ÷ Ⓑ = 7 in.
Ⓓ . . . 11 – Ⓒ = 4 in.
Ⓔ . . . 10 – Ⓓ = 6 in.
Ⓕ . . . 30 ÷ Ⓔ = 5 in.
Ⓖ . . . 12 – Ⓕ = 7 in.
Area ? is Ⓖ × 4 = 28 in.²

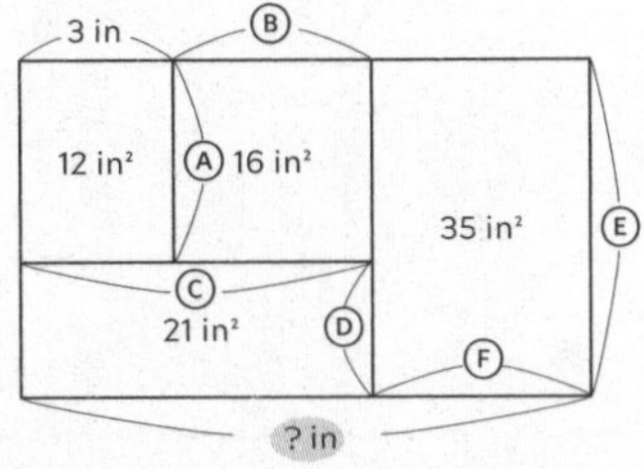

Puzzle 10

Ⓐ . . . 12 ÷ 3 = 4 in.
Ⓑ . . . 16 ÷ Ⓐ = 4 in.
Ⓒ . . . 3 + Ⓑ = 7 in.
Ⓓ . . . 21 ÷ Ⓒ = 3 in.
Ⓔ . . . Ⓐ + Ⓓ = 7 in.
Ⓕ . . . 35 ÷ Ⓔ = 5 in.
Length ? is Ⓒ + Ⓕ = 12 in.

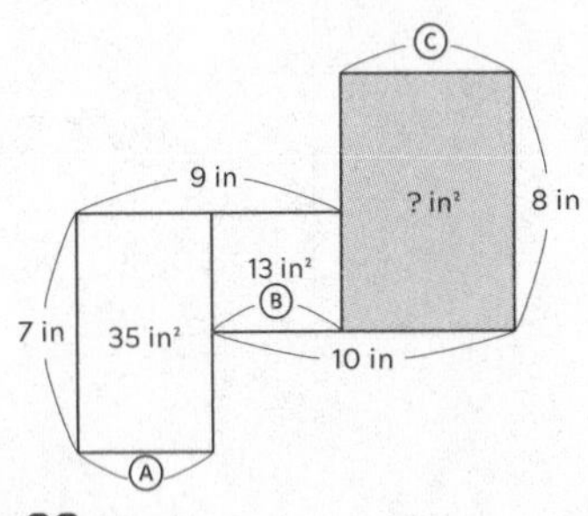

Puzzle 11

Ⓐ . . . 35 ÷ 7 = 5 in.
Ⓑ . . . 9 – Ⓐ = 4 in.
Ⓒ . . . 10 – Ⓑ = 6 in.
Area ? is Ⓒ × 8 = 48 in.²

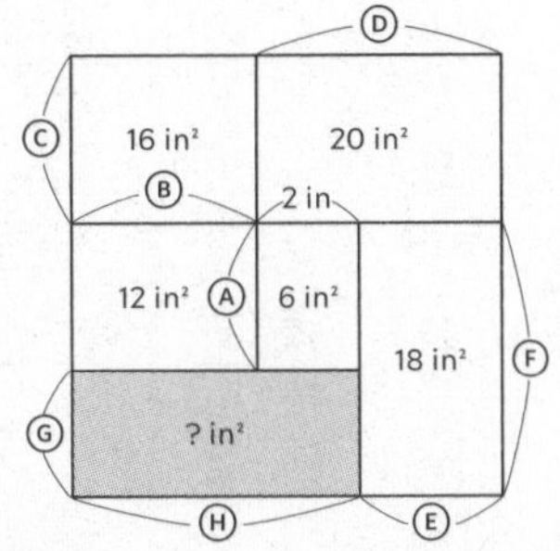

Puzzle 12

Ⓐ . . . 6 ÷ 2 = 3 in.
Ⓑ . . . 12 ÷ Ⓐ = 4 in.
Ⓒ . . . 16 ÷ Ⓑ = 4 in.
Ⓓ . . . 20 ÷ Ⓒ = 5 in.
Ⓔ . . . Ⓓ – 2 = 3 in.
Ⓕ . . . 18 ÷ Ⓔ = 6 in.
Ⓖ . . . Ⓕ – Ⓐ = 3 in.
Ⓗ . . . Ⓑ + 2 = 6 in.
Area ? is Ⓖ × Ⓗ = 18 in.²

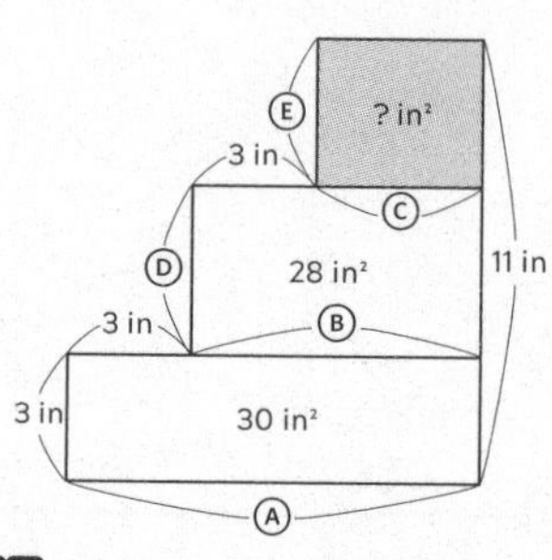

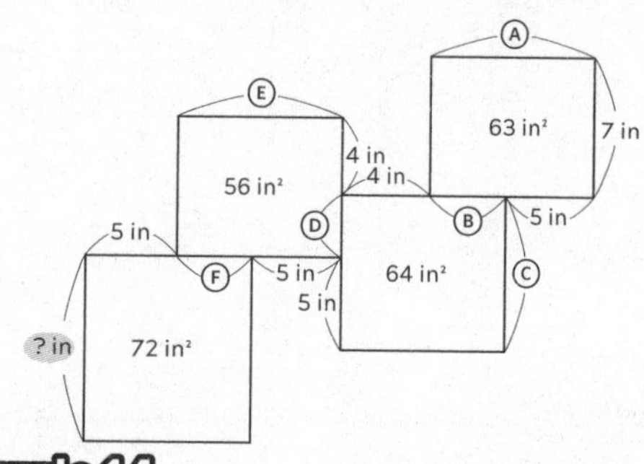

Puzzle 13

Ⓐ . . . 30 ÷ 3 = 10 in.
Ⓑ . . . Ⓐ – 3 = 7 in.
Ⓒ . . . Ⓑ – 3 = 4 in.
Ⓓ . . . 28 ÷ Ⓑ = 4 in.
Ⓔ . . . 11 – 3 – Ⓓ = 4 in.
Area ? is Ⓒ × Ⓔ = 16 in.²

Puzzle 14

Ⓐ . . . 63 ÷ 7 = 9 in.
Ⓑ . . . Ⓐ – 5 = 4 in.
Ⓒ . . . 64 ÷ (Ⓑ + 4) = 8 in.
Ⓓ . . . Ⓒ – 5 = 3 in.
Ⓔ . . . 56 ÷ (Ⓓ + 4) = 8 in.
Ⓕ . . . Ⓔ – 5 = 3 in.
Length ? is 72 ÷ (5 + Ⓕ) = 9 in.

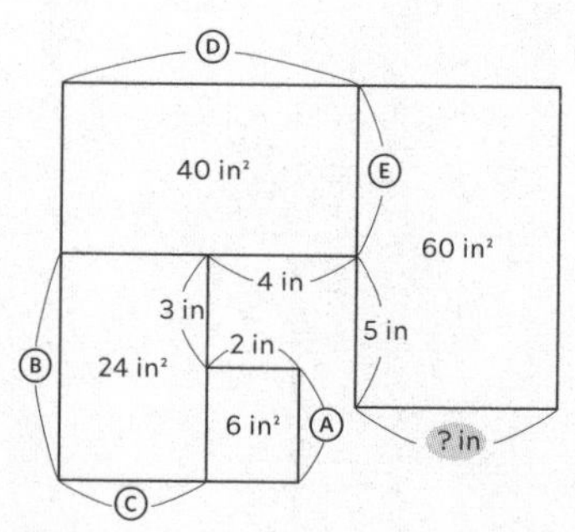

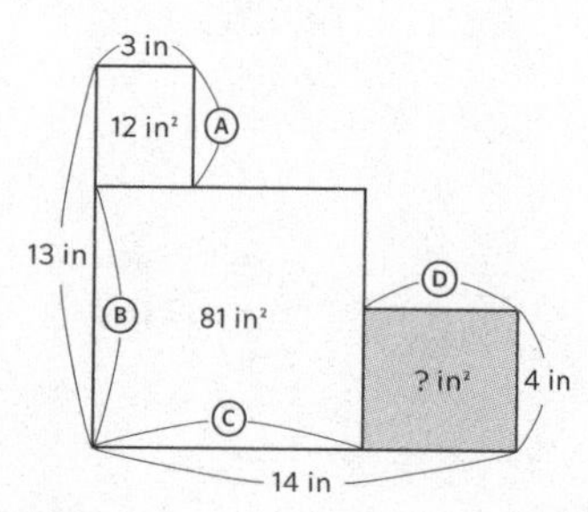

Puzzle 15

Ⓐ . . . 6 ÷ 2 = 3 in.
Ⓑ . . . Ⓐ + 3 = 6 in.
Ⓒ . . . 24 ÷ Ⓑ = 4 in.
Ⓓ . . . Ⓒ + 4 = 8 in.
Ⓔ . . . 40 ÷ Ⓓ = 5 in.
Length ? is 60 ÷ (Ⓔ + 5) = 6 in.

Puzzle 16

Ⓐ . . . 12 ÷ 3 = 4 in.
Ⓑ . . . 13 – Ⓐ = 9 in.
Ⓒ . . . 81 ÷ Ⓑ = 9 in.
Ⓓ . . . 14 – Ⓒ = 5 in.
Area ? is Ⓓ × 4 = 20 in.²

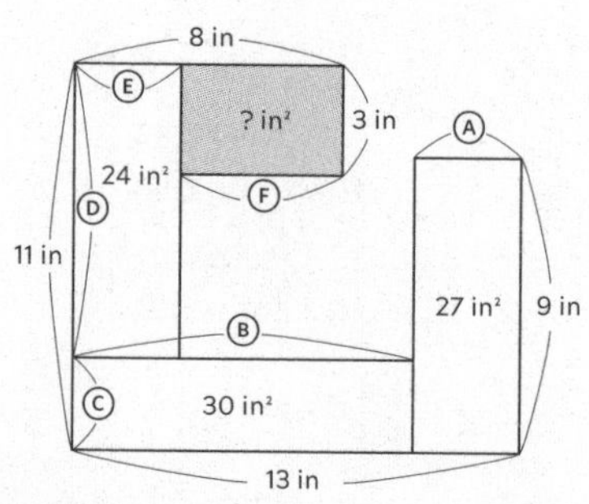

Puzzle 17

Ⓐ . . . 27 ÷ 9 = 3 in.
Ⓑ . . . 13 – Ⓐ = 10 in.
Ⓒ . . . 30 ÷ Ⓑ = 3 in.
Ⓓ . . . 11 – Ⓒ = 8 in.
Ⓔ . . . 24 ÷ Ⓓ = 3 in.
Ⓕ . . . 8 – Ⓔ = 5 in.
Area ? is Ⓕ × 3 = 15 in.²

Puzzle 18

Ⓐ . . . 18 ÷ 3 = 6 in.
Ⓑ . . . 10 – Ⓐ = 4 in.
Ⓒ . . . 20 ÷ Ⓑ = 5 in.
Ⓓ . . . Ⓒ + 5 – 3 = 7 in.
Ⓔ . . . 28 ÷ Ⓓ = 4 in.
Ⓕ . . . 35 ÷ 5 = 7 in.
Length ? is Ⓔ + Ⓕ = 11 in.

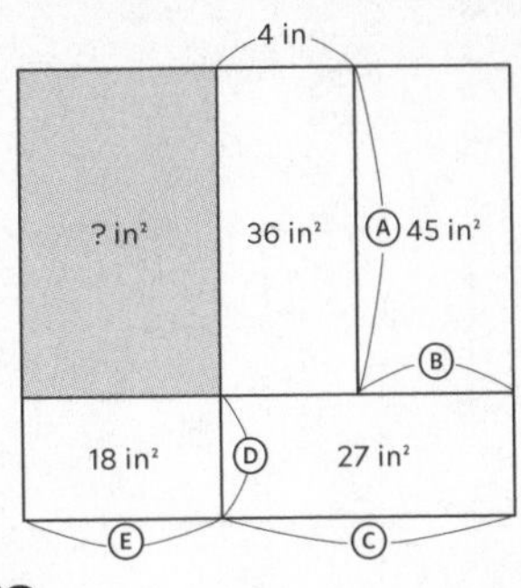

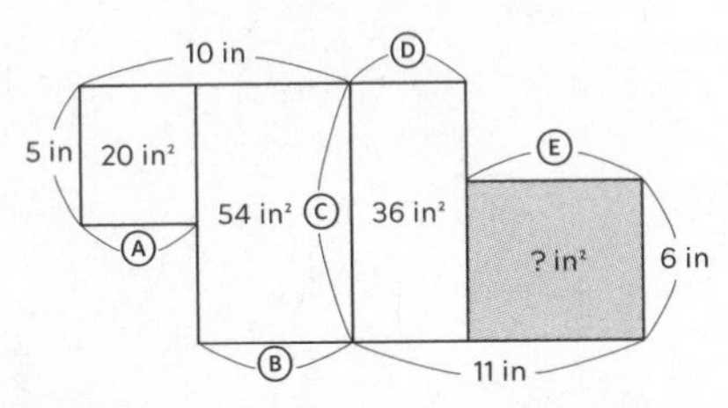

Puzzle 19

Ⓐ . . . 36 ÷ 4 = 9 in.
Ⓑ . . . 45 ÷ Ⓐ = 5 in.
Ⓒ . . . 4 + Ⓑ = 9 in.
Ⓓ . . . 27 ÷ Ⓒ = 3 in.
Ⓔ . . . 18 ÷ Ⓓ = 6 in.
Area ? is Ⓐ × Ⓔ = 54 in.²

Puzzle 20

Ⓐ . . . 20 ÷ 5 = 4 in.
Ⓑ . . . 10 – Ⓐ = 6 in.
Ⓒ . . . 54 ÷ Ⓑ = 9 in.
Ⓓ . . . 36 ÷ Ⓒ = 4 in.
Ⓔ . . . 11 – Ⓓ = 7 in.
Area ? is Ⓔ × 6 = 42 in.²

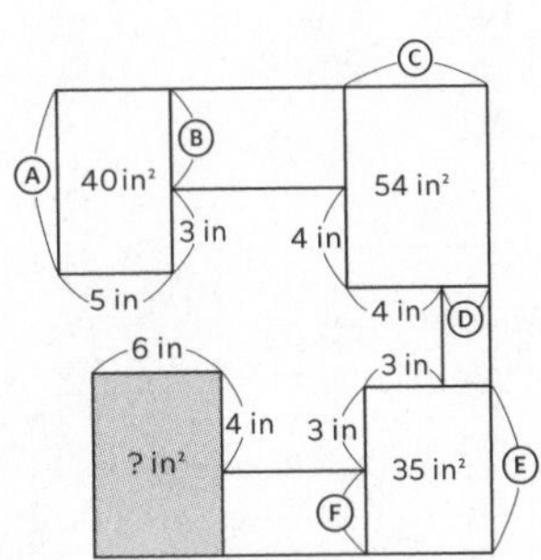

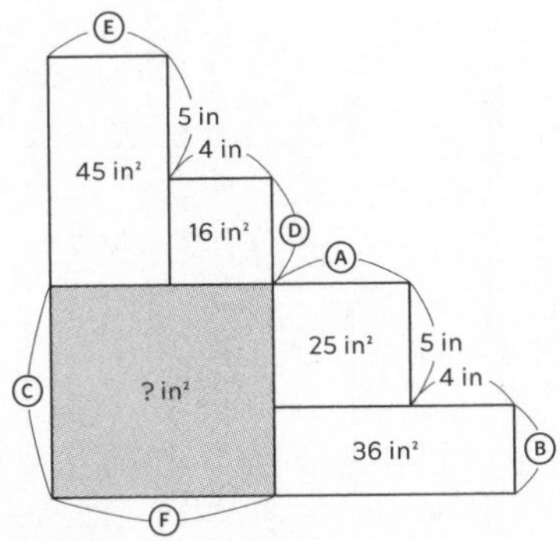

Puzzle 21

Ⓐ . . . 40 ÷ 5 = 8 in.
Ⓑ . . . Ⓐ – 3 = 5 in.
Ⓒ . . . 54 ÷ (Ⓑ + 4) = 6 in.
Ⓓ . . . Ⓒ – 4 = 2 in.
Ⓔ . . . 35 ÷ (Ⓓ + 3) = 7 in.
Ⓕ . . . Ⓔ – 3 = 4 in.
Area ? is (Ⓕ + 4) × 6
= 48 in.²

Puzzle 22

Ⓐ . . . 25 ÷ 5 = 5 in.
Ⓑ . . . 36 ÷ (Ⓐ + 4) = 4 in.
Ⓒ . . . 5 + Ⓑ = 9 in.
Ⓓ . . . 16 ÷ 4 = 4 in.
Ⓔ . . . 45 ÷ (5 + Ⓓ) = 5 in.
Ⓕ . . . Ⓔ + 4 = 9 in.
Area ? is Ⓒ × Ⓕ = 81 in.²

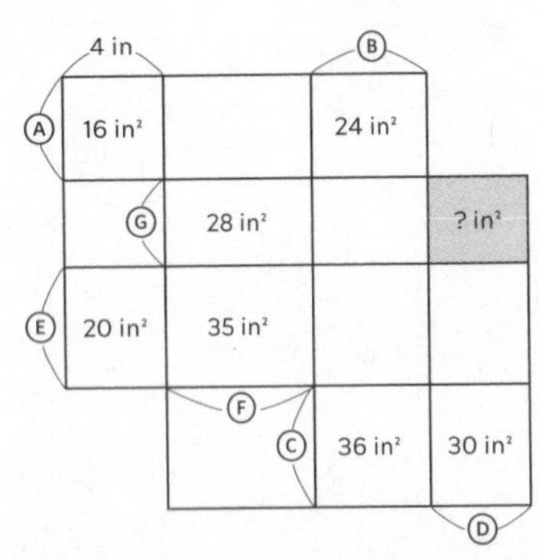

Puzzle 23

Ⓐ . . . 16 ÷ 4 = 4 in.
Ⓑ . . . 24 ÷ Ⓐ = 6 in.
Ⓒ . . . 36 ÷ Ⓑ = 6 in.
Ⓓ . . . 30 ÷ Ⓒ = 5 in.
Ⓔ . . . 20 ÷ 4 = 5 in.
Ⓕ . . . 35 ÷ Ⓔ = 7 in.
Ⓖ . . . 28 ÷ Ⓕ = 4 in.
Area ? is Ⓓ × Ⓖ = 20 in.²

Puzzle 24

Ⓐ . . . 10 – 2 = 8 in.
Ⓑ . . . 48 ÷ Ⓐ = 6 in.
Ⓒ . . . 42 ÷ Ⓑ = 7 in.
Ⓓ . . . 56 ÷ Ⓒ = 8 in.
Area ? is 10 × (Ⓓ + 2)
= 100 in.²

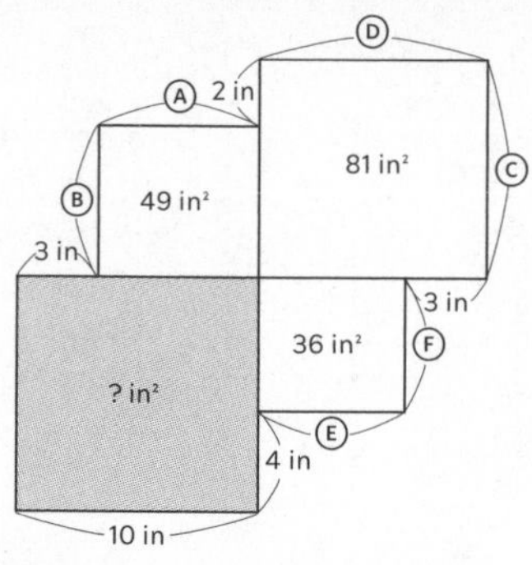

Puzzle 25

Ⓐ . . . 10 − 3 = 7 in.
Ⓑ . . . 49 ÷ Ⓐ = 7 in.
Ⓒ . . . Ⓑ + 2 = 9 in.
Ⓓ . . . 81 ÷ Ⓒ = 9 in.
Ⓔ . . . Ⓓ − 3 = 6 in.
Ⓕ . . . 36 ÷ Ⓔ = 6 in.
Area ? is (Ⓕ + 4) × 10
= 100 in.^{2}

Puzzle 26

Ⓐ . . . 48 ÷ 8 = 6 in.
Ⓑ . . . 9 − Ⓐ = 3 in.
Ⓒ . . . 40 ÷ 10 = 4 in.
Ⓓ . . . 8 − Ⓒ = 4 in.
Area ? is Ⓑ × Ⓓ = 12 in.^{2}

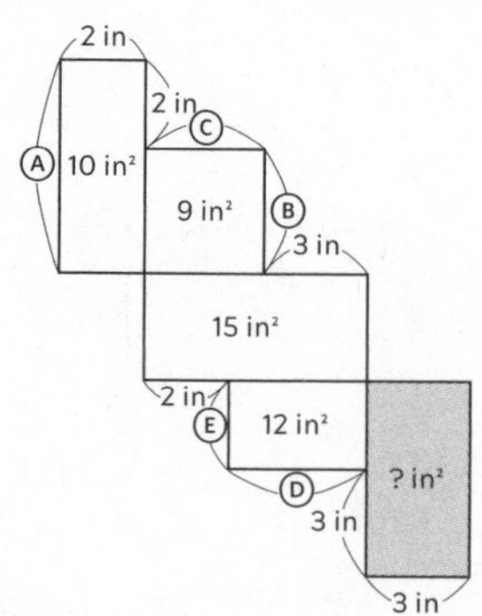

Puzzle 27

Ⓐ . . . 10 ÷ 2 = 5 in.
Ⓑ . . . Ⓐ − 2 = 3 in.
Ⓒ . . . 9 ÷ Ⓑ = 3 in.
Ⓓ . . . Ⓒ + 3 − 2 = 4 in.
Ⓔ . . . 12 ÷ Ⓓ = 3 in.
Area ? is (Ⓔ + 3) × 3
= 18 in.^{2}

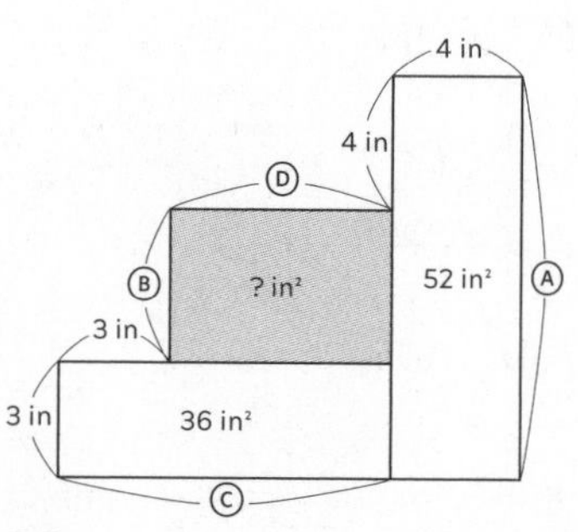

Puzzle 28

Ⓐ . . . 52 ÷ 4 = 13 in.
Ⓑ . . . Ⓐ − 4 − 3 = 6 in.
Ⓒ . . . 36 ÷ 3 = 12 in.
Ⓓ . . . Ⓒ − 3 = 9 in.
Area ? is Ⓑ × Ⓓ = 54 in.^{2}

Puzzle 29

Ⓐ . . . 3 + 5 = 8 in.
Ⓑ . . . 32 ÷ Ⓐ = 4 in.
Ⓒ . . . 4 + Ⓑ = 8 in.
Ⓓ . . . 24 ÷ Ⓒ = 3 in.
Ⓔ . . . Ⓓ + 6 = 9 in.
Ⓕ . . . 36 ÷ Ⓔ = 4 in.
Area ? is 5 × (Ⓕ + 5)
= 45 in.^{2}

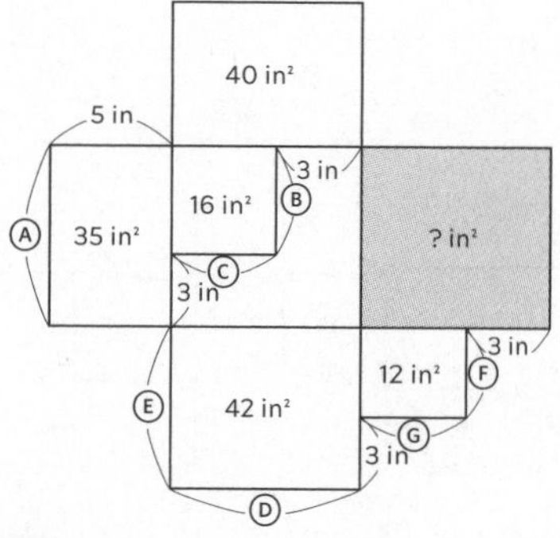

Puzzle 30

Ⓐ . . . 35 ÷ 5 = 7 in.
Ⓑ . . . Ⓐ − 3 = 4 in.
Ⓒ . . . 16 ÷ Ⓑ = 4 in.
Ⓓ . . . Ⓒ + 3 = 7 in.
Ⓔ . . . 42 ÷ Ⓓ = 6 in.
Ⓕ . . . Ⓔ − 3 = 3 in.
Ⓖ . . . 12 ÷ Ⓕ = 4 in.
Area ? is Ⓐ × (Ⓖ + 3)
= 49 in.^{2}

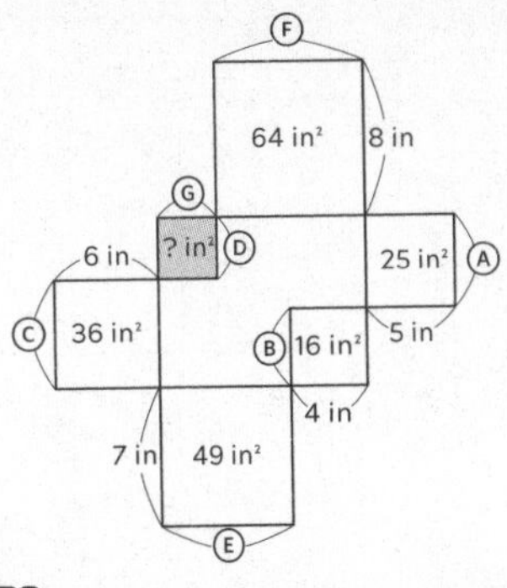

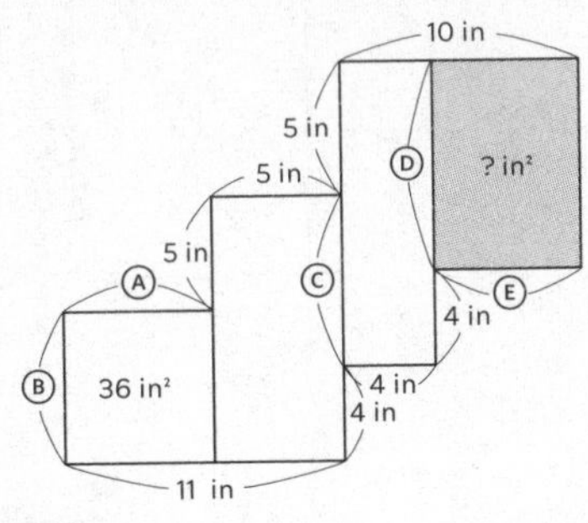

Puzzle 31

Ⓐ . . . 25 ÷ 5 = 5 in.
Ⓑ . . . 16 ÷ 4 = 4 in.
Ⓒ . . . 36 ÷ 6 = 6 in.
Ⓓ . . . Ⓐ + Ⓑ − Ⓒ = 3 in.
Ⓔ . . . 49 ÷ 7 = 7 in.
Ⓕ . . . 64 ÷ 8 = 8 in.
Ⓖ . . . Ⓔ + 4 − Ⓕ = 3 in.
Area ? is Ⓓ × Ⓖ = 9 in.^{2}

Puzzle 32

Ⓐ . . . 11 − 5 = 6 in.
Ⓑ . . . 36 ÷ Ⓐ = 6 in.
Ⓒ . . . Ⓑ + 5 − 4 = 7 in.
Ⓓ . . . 5 + Ⓒ − 4 = 8 in.
Ⓔ . . . 10 − 4 = 6 in.
Area ? is Ⓓ × Ⓔ = 48 in.^{2}

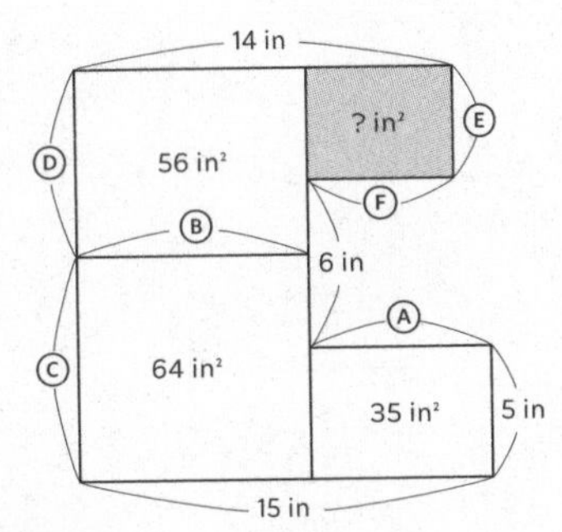

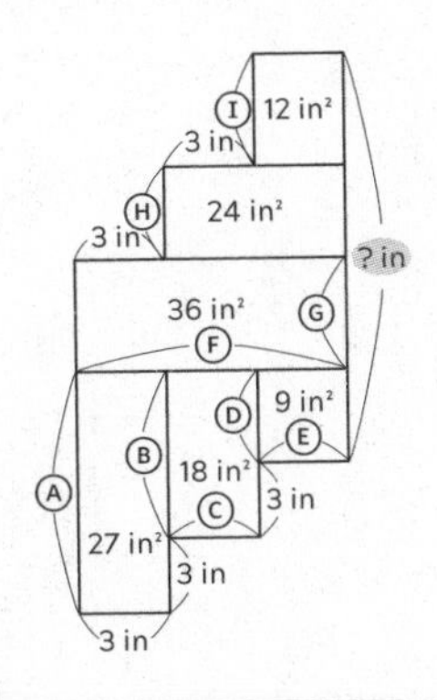

Puzzle 33

Ⓐ . . . 35 ÷ 5 = 7 in.
Ⓑ . . . 15 − Ⓐ = 8 in.
Ⓒ . . . 64 ÷ Ⓑ = 8 in.
Ⓓ . . . 56 ÷ Ⓑ = 7 in.
Ⓔ . . . Ⓒ + Ⓓ − 5 − 6 = 4 in.
Ⓕ . . . 14 − Ⓑ = 6 in.
Area ? is Ⓔ × Ⓕ = 24 in.^{2}

Puzzle 34

Ⓐ . . . 27 ÷ 3 = 9 in.
Ⓑ . . . Ⓐ − 3 = 6 in.
Ⓒ . . . 18 ÷ Ⓑ = 3 in.
Ⓓ . . . Ⓑ − 3 = 3 in.
Ⓔ . . . 9 ÷ Ⓓ = 3 in.
Ⓕ . . . 3 + Ⓒ + Ⓔ = 9 in.
Ⓖ . . . 36 ÷ Ⓕ = 4 in.
Ⓗ . . . 24 ÷ (Ⓕ − 3) = 4 in.
Ⓘ . . . 12 ÷ (Ⓕ − 3 − 3) = 4 in.
Length ? is Ⓓ + Ⓖ + Ⓗ + Ⓘ = 15 in.

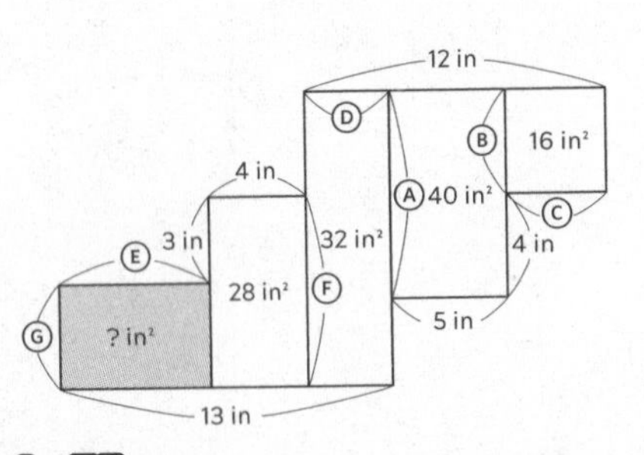

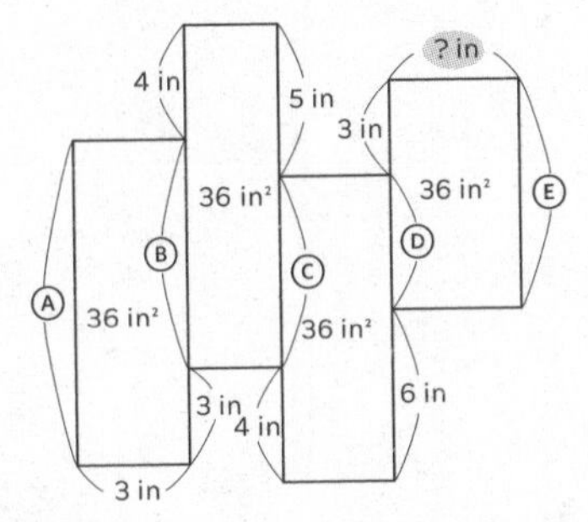

Puzzle 35

Ⓐ . . . 40 ÷ 5 = 8 in.
Ⓑ . . . Ⓐ − 4 = 4 in.
Ⓒ . . . 16 ÷ Ⓑ = 4 in.
Ⓓ . . . 12 − 5 − Ⓒ = 3 in.
Ⓔ . . . 13 − Ⓓ − 4 = 6 in.
Ⓕ . . . 28 ÷ 4 = 7 in.
Ⓖ . . . Ⓕ − 3 = 4 in.
Area ? is Ⓔ × Ⓖ = 24 in.^{2}

Puzzle 36

Ⓐ . . . 36 ÷ 3 = 12 in.
Ⓑ . . . Ⓐ − 3 = 9 in.
Ⓒ . . . Ⓑ + 4 − 5 = 8 in.
Ⓓ . . . Ⓒ + 4 − 6 = 6 in.
Ⓔ . . . Ⓓ + 3 = 9 in.
Length ? is 36 ÷ Ⓔ = 4 in.

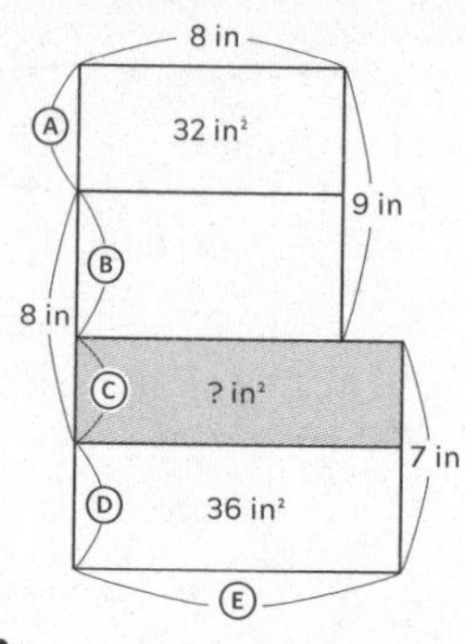

Puzzle 37

Ⓐ . . . 32 ÷ 8 = 4 in.
Ⓑ . . . 9 − Ⓐ = 5 in.
Ⓒ . . . 8 − Ⓑ = 3 in.
Ⓓ . . . 7 − Ⓒ = 4 in.
Ⓔ . . . 36 ÷ Ⓓ = 9 in.
Area ? is Ⓒ × Ⓔ = 27 in.^{2}

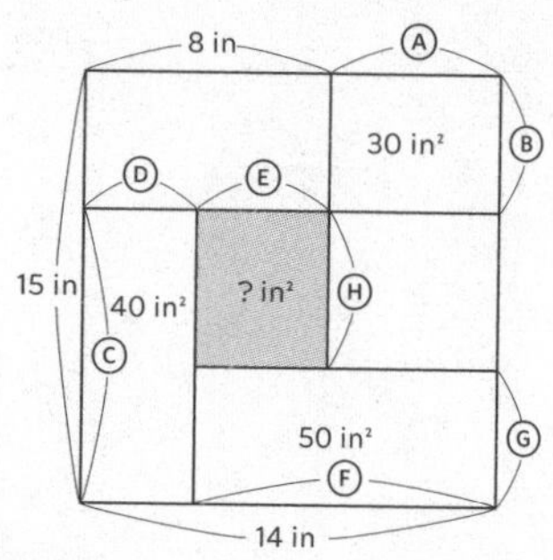

Puzzle 38

Ⓐ . . . 14 − 8 = 6 in.
Ⓑ . . . 30 ÷ Ⓐ = 5 in.
Ⓒ . . . 15 − Ⓑ = 10 in.
Ⓓ . . . 40 ÷ Ⓒ = 4 in.
Ⓔ . . . 8 − Ⓓ = 4 in.
Ⓕ . . . 14 − Ⓓ = 10 in.
Ⓖ . . . 50 ÷ Ⓕ = 5 in.
Ⓗ . . . Ⓒ − Ⓖ = 5 in.
Area ? is Ⓔ × Ⓗ = 20 in.^{2}

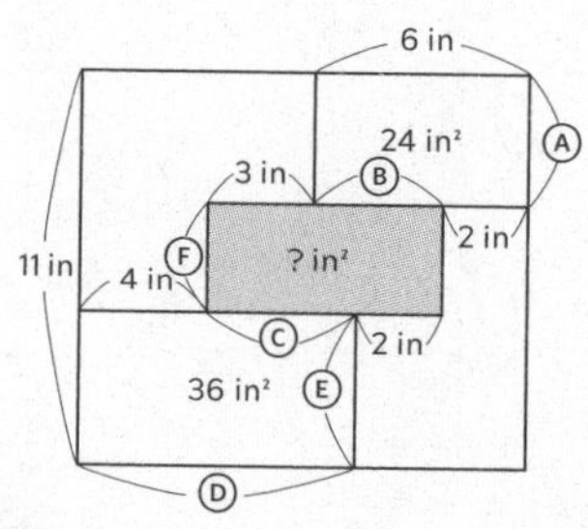

Puzzle 39

Ⓐ . . . 24 ÷ 6 = 4 in.
Ⓑ . . . 6 − 2 = 4 in.
Ⓒ . . . Ⓑ + 3 − 2 = 5 in.
Ⓓ . . . 4 + Ⓒ = 9 in.
Ⓔ . . . 36 ÷ Ⓓ = 4 in.
Ⓕ . . . 11 − Ⓐ − Ⓔ = 3 in.
Area ? is (Ⓑ + 3) × Ⓕ
= 21 in.^{2}

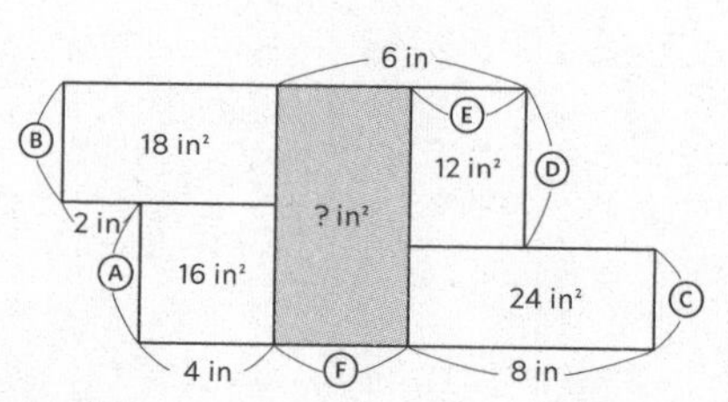

Puzzle 40

Ⓐ . . . 16 ÷ 4 = 4 in.
Ⓑ . . . 18 ÷ (2 + 4) = 3 in.
Ⓒ . . . 24 ÷ 8 = 3 in.
Ⓓ . . . Ⓐ + Ⓑ − Ⓒ = 4 in.
Ⓔ . . . 12 ÷ Ⓓ = 3 in.
Ⓕ . . . 6 − Ⓔ = 3 in.
Area ? is (Ⓐ + Ⓑ) × Ⓕ
= 21 in.^{2}

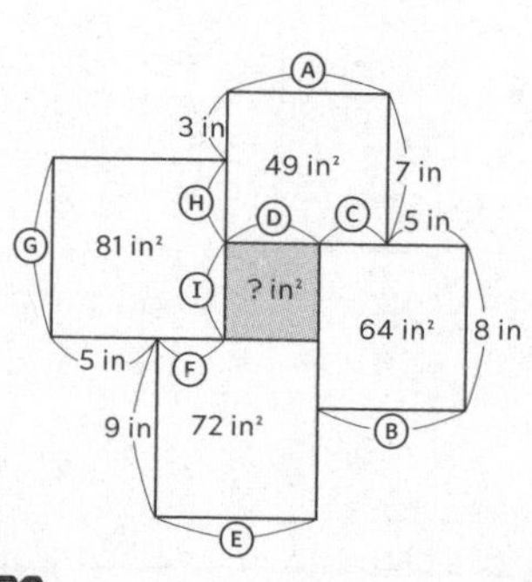

Puzzle 41

Ⓐ . . . 49 ÷ 7 = 7 in.
Ⓑ . . . 64 ÷ 8 = 8 in.
Ⓒ . . . Ⓑ − 5 = 3 in.
Ⓓ . . . Ⓐ − Ⓒ = 4 in.
Ⓔ . . . 72 ÷ 9 = 8 in.
Ⓕ . . . Ⓔ − Ⓓ = 4 in.
Ⓖ . . . 81 ÷ (5 + Ⓕ) = 9 in.
Ⓗ . . . 7 − 3 = 4 in.
Ⓘ . . . Ⓖ − Ⓗ = 5 in.
Area ? is Ⓓ × Ⓘ = 20 in.^{2}

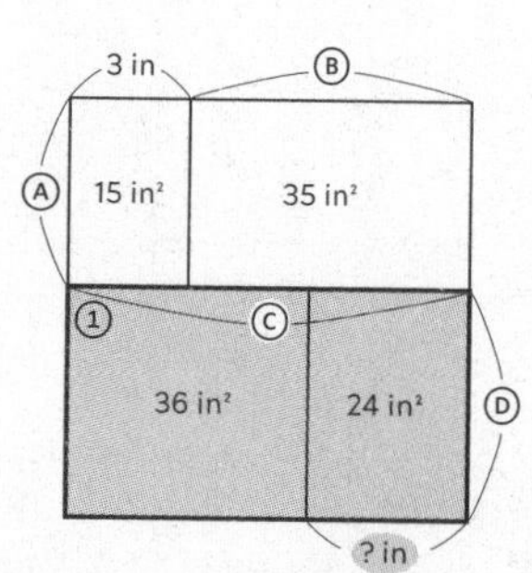

Puzzle 42

Ⓐ . . . 15 ÷ 3 = 5 in.
Ⓑ . . . 35 ÷ Ⓐ = 7 in.
Ⓒ . . . 3 + Ⓑ = 10 in.
Area ① . . .
36 + 24 = 60 in.^{2}
Ⓓ . . . 60 ÷ Ⓒ = 6 in.
Length ? is 24 ÷ Ⓓ = 4 in.

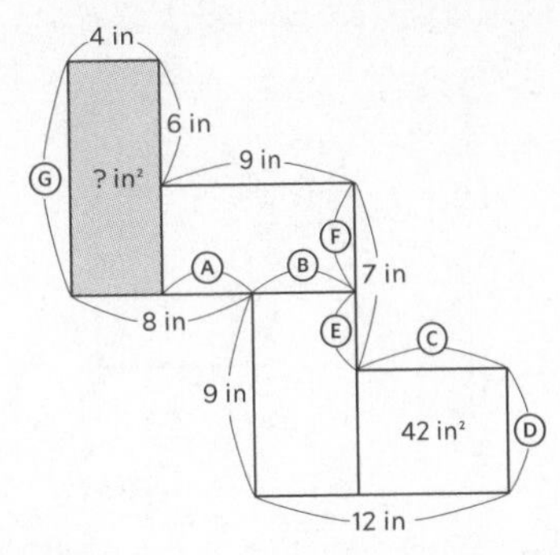

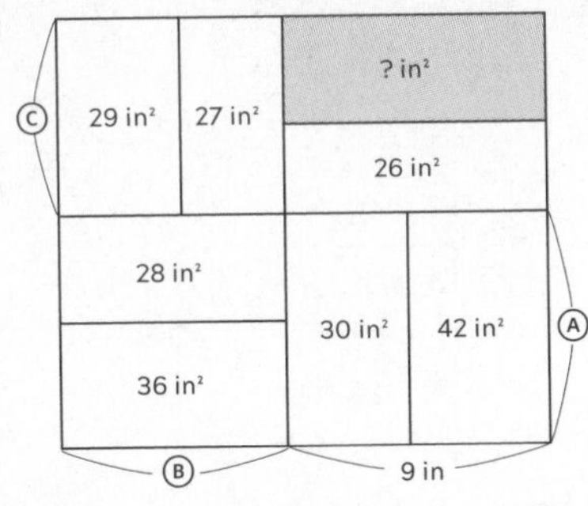

Puzzle 43

Ⓐ . . . 8 – 4 = 4 in.
Ⓑ . . . 9 – Ⓐ = 5 in.
Ⓒ . . . 12 – Ⓑ = 7 in.
Ⓓ . . . 42 ÷ Ⓒ = 6 in.
Ⓔ . . . 9 – Ⓓ = 3 in.
Ⓕ . . . 7 – Ⓔ = 4 in.
Ⓖ . . . 6 + Ⓕ = 10 in.
Area (?) is Ⓖ × 4 = 40 in.^{2}

Puzzle 44

Ⓐ . . . (30 + 42) ÷ 9 = 8 in.
Ⓑ . . . (28 + 36) ÷ Ⓐ = 8 in.
Ⓒ . . . (29 + 27) ÷ Ⓑ = 7 in.
Area (?) is (9 × Ⓒ) – 26 = 37 in.^{2}

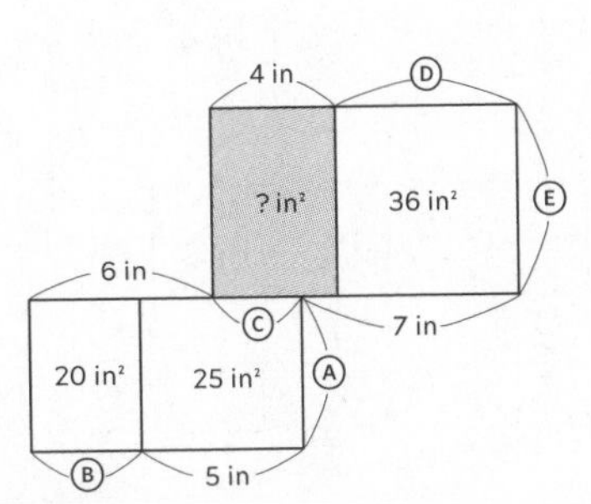

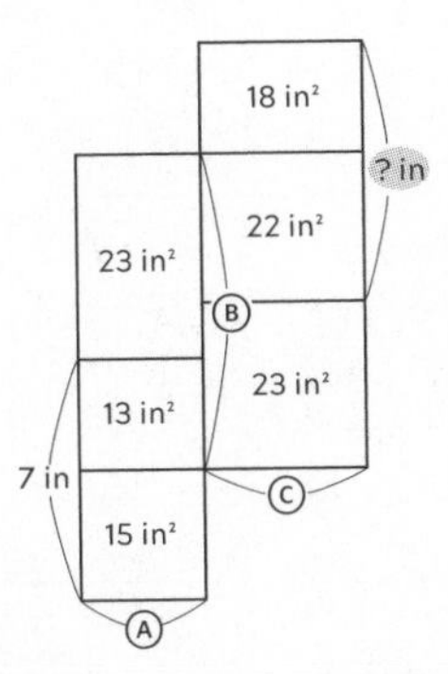

Puzzle 45

Ⓐ . . . 25 ÷ 5 = 5 in.
Ⓑ . . . 20 ÷ Ⓐ = 4 in.
Ⓒ . . . Ⓑ + 5 – 6 = 3 in.
Ⓓ . . . Ⓒ + 7 – 4 = 6 in.
Ⓔ . . . 36 ÷ Ⓓ = 6 in.
Area (?) is 4 × Ⓔ = 24 in.^{2}

Puzzle 46

Ⓐ . . . (13 + 15) ÷ 7 = 4 in.
Ⓑ . . . (23 + 13) ÷ Ⓐ = 9 in.
Ⓒ . . . (22 + 23) ÷ Ⓑ = 5 in.
Length (?) is (18 + 22) ÷ Ⓒ = 8 in.

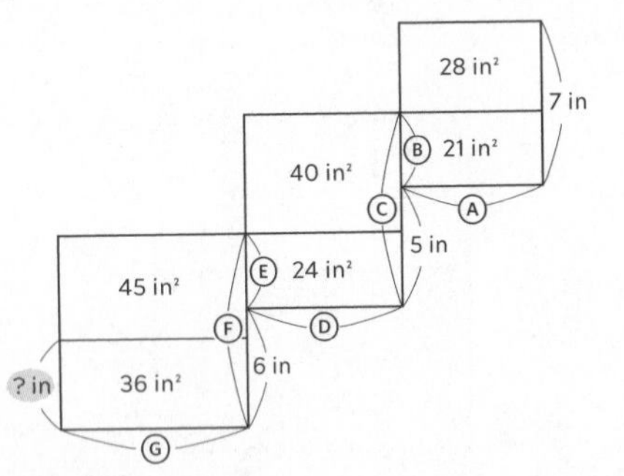

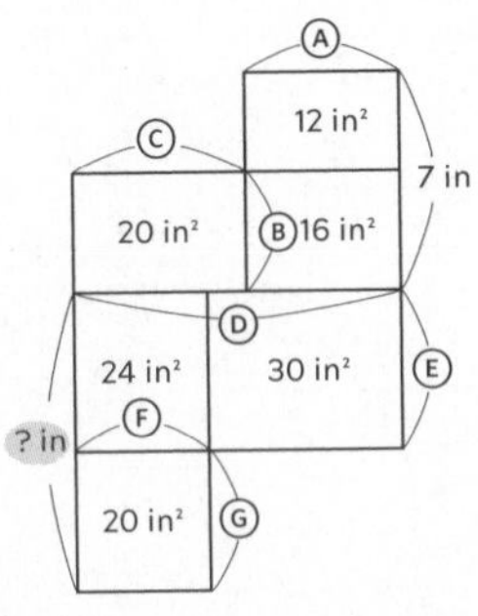

Puzzle 47

Ⓐ . . . (28 + 21) ÷ 7 = 7 in.
Ⓑ . . . 21 ÷ Ⓐ = 3 in.
Ⓒ . . . Ⓑ + 5 = 8 in.
Ⓓ . . . (40 + 24) ÷ Ⓒ = 8 in.
Ⓔ . . . 24 ÷ Ⓓ = 3 in.
Ⓕ . . . Ⓔ + 6 = 9 in.
Ⓖ . . . (45 + 36) ÷ Ⓕ = 9 in.
Length (?) is 36 ÷ Ⓖ = 4 in.

Puzzle 48

Ⓐ . . . (12 + 16) ÷ 7 = 4 in.
Ⓑ . . . 16 ÷ Ⓐ = 4 in.
Ⓒ . . . 20 ÷ Ⓑ = 5 in.
Ⓓ . . . Ⓐ + Ⓒ = 9 in.
Ⓔ . . . (24 + 30) ÷ Ⓓ = 6 in.
Ⓕ . . . 24 ÷ Ⓔ = 4 in.
Ⓖ . . . 20 ÷ Ⓕ = 5 in.
Length (?) is Ⓔ + Ⓖ = 11 in.

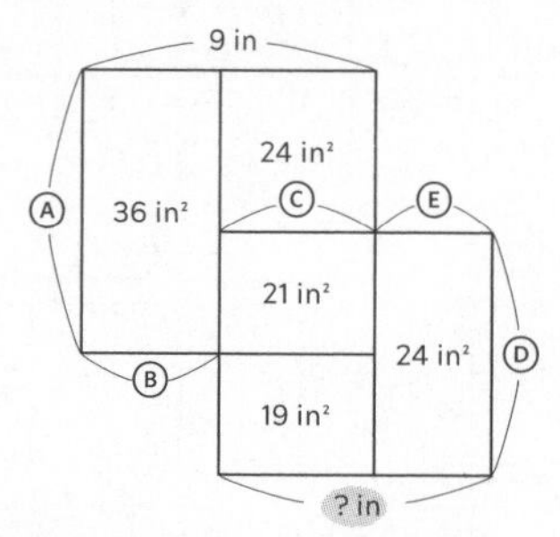

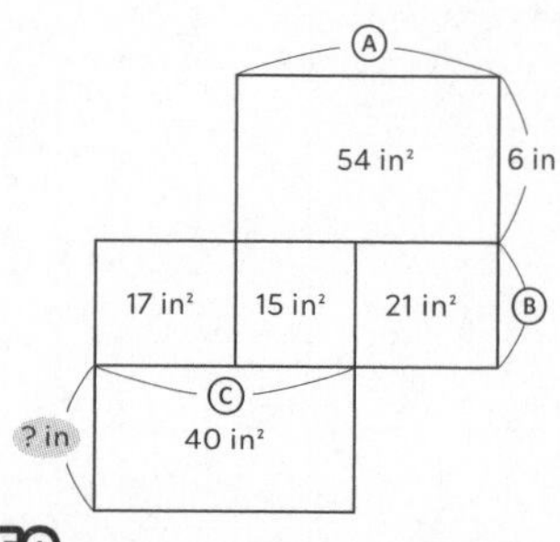

Puzzle 49

Ⓐ . . . (36 + 24 + 21) ÷ 9
= 9 in.
Ⓑ . . . 36 ÷ Ⓐ = 4 in.
Ⓒ . . . 9 − Ⓑ = 5 in.

Ⓓ . . . (21 + 19) ÷ Ⓒ = 8 in.
Ⓔ . . . 24 ÷ Ⓓ = 3 in.
Length ? is Ⓒ + Ⓔ = 8 in.

Puzzle 50

Ⓐ . . . 54 ÷ 6 = 9 in.
Ⓑ . . . (15 + 21) ÷ Ⓐ = 4 in.

Ⓒ . . . (17 + 15) ÷ Ⓑ = 8 in.
Length ? is 40 ÷ Ⓒ = 5 in.

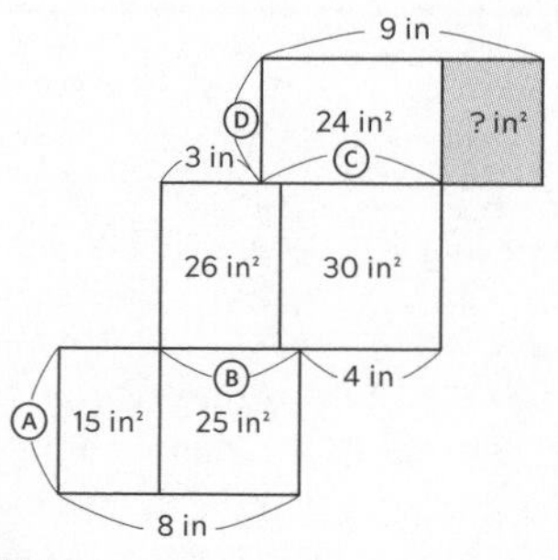

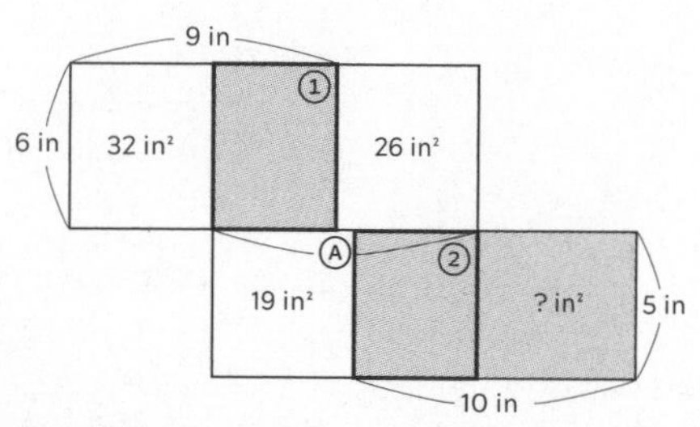

Puzzle 51

Ⓐ . . . (15 + 25) ÷ 8 = 5 in.
Ⓑ . . . 25 ÷ Ⓐ = 5 in.
Ⓒ . . . Ⓑ + 4 − 3 = 6 in.

Ⓓ . . . 24 ÷ Ⓒ = 4 in.
Area ? is (9 - Ⓒ) × Ⓓ
= 12 in.²

Puzzle 52

Area ① . . .
(6 × 9) − 32 = 22 in.²
Ⓐ . . . (22 + 26) ÷ 6 = 8 in.

Area ② . . .
(Ⓐ × 5) − 19 = 21 in.²
Area ? is (5 × 10) − 21
= 29 in.²

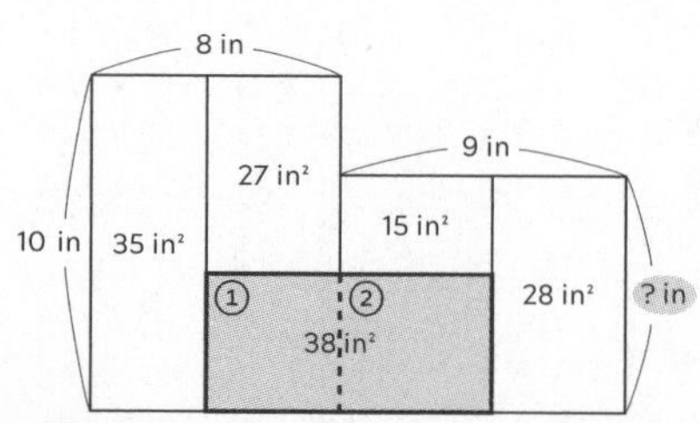

Puzzle 53

Area ① . . .
(10 × 8) − 35 − 27
= 18 in.²
Area ② . . . 38 − 18 = 20 in.²

Length ? is
(15 + 20 + 28) ÷ 9
= 7 in.

Puzzle 54

Ⓐ . . . 20 ÷ 4 = 5 in.
Ⓑ . . . (18 + 20 + 17) ÷ Ⓐ
= 11 in.
Ⓒ . . . Ⓑ − 2 − 2 = 7 in.

Ⓓ . . . (24 + 18) ÷ Ⓒ = 6 in.
Ⓔ . . . (24 ÷ Ⓓ) − 2 = 2 in.
Length ? is Ⓔ + 2 = 4 in.

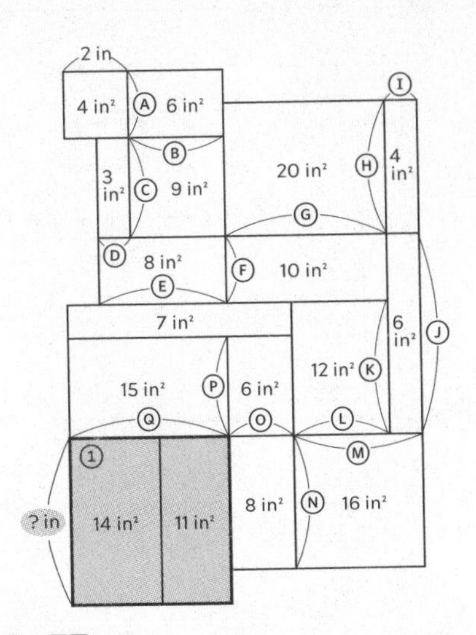

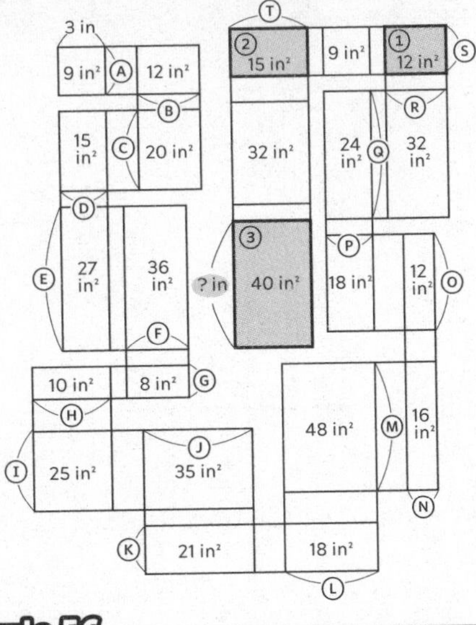

Puzzle 55

Ⓐ . . . 4 ÷ 2 = 2 in.
Ⓑ . . . 6 ÷ Ⓐ = 3 in.
Ⓒ . . . 9 ÷ Ⓑ = 3 in.
Ⓓ . . . 3 ÷ Ⓒ = 1 in.
Ⓔ . . . Ⓑ + Ⓓ = 4 in.
Ⓕ . . . 8 ÷ Ⓔ = 2 in.
Ⓖ . . . 10 ÷ Ⓕ = 5 in.
Ⓗ . . . 20 ÷ Ⓖ = 4 in.
Ⓘ . . . 4 ÷ Ⓗ = 1 in.
Ⓙ . . . 6 ÷ Ⓘ = 6 in.
Ⓚ . . . Ⓙ − Ⓕ = 4 in.
Ⓛ . . . 12 ÷ Ⓚ = 3 in.
Ⓜ . . . Ⓘ + Ⓛ = 4 in.
Ⓝ . . . 16 ÷ Ⓜ = 4 in.
Ⓞ . . . 8 ÷ Ⓝ = 2 in.
Ⓟ . . . 6 ÷ Ⓞ = 3 in.
Ⓠ . . . 15 ÷ Ⓟ = 5 in.
Area ① . . .
14 + 11 = 25 in.^{2}
Length ⓘ is 25 ÷ Ⓠ = 5 in.

Puzzle 56

Ⓐ . . . 9 ÷ 3 = 3 in.
Ⓑ . . . 12 ÷ Ⓐ = 4 in.
Ⓒ . . . 20 ÷ Ⓑ = 5 in.
Ⓓ . . . 15 ÷ Ⓒ = 3 in.
Ⓔ . . . 27 ÷ Ⓓ = 9 in.
Ⓕ . . . 36 ÷ Ⓔ = 4 in.
Ⓖ . . . 8 ÷ Ⓕ = 2 in.
Ⓗ . . . 10 ÷ Ⓖ = 5 in.
Ⓘ . . . 25 ÷ Ⓗ = 5 in.
Ⓙ . . . 35 ÷ Ⓘ = 7 in.
Ⓚ . . . 21 ÷ Ⓙ = 3 in.
Ⓛ . . . 18 ÷ Ⓚ = 6 in.
Ⓜ . . . 48 ÷ Ⓛ = 8 in.
Ⓝ . . . 16 ÷ Ⓜ = 2 in.
Ⓞ . . . 12 ÷ Ⓝ = 6 in.
Ⓟ . . . 18 ÷ Ⓞ = 3 in.
Ⓠ . . . 24 ÷ Ⓟ = 8 in.
Ⓡ . . . 32 ÷ Ⓠ = 4 in.
Ⓢ . . . 12 ÷ Ⓡ = 3 in.
Areas ① and ② are the same height.
Ⓣ . . . 15 ÷ Ⓢ = 5 in.
Areas ② and ③ are the same width.
Length ⓘ is 40 ÷ Ⓣ = 8 in.

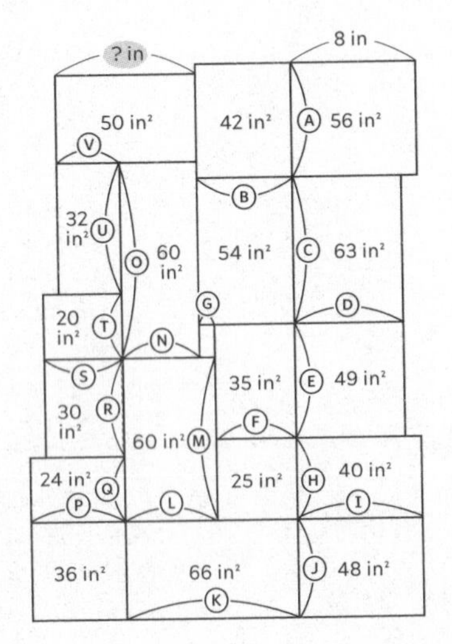

Puzzle 57

Ⓐ . . . 56 ÷ 8 = 7 in.
Ⓑ . . . 42 ÷ Ⓐ = 6 in.
Ⓒ . . . 54 ÷ Ⓑ = 9 in.
Ⓓ . . . 63 ÷ Ⓒ = 7 in.
Ⓔ . . . 49 ÷ Ⓓ = 7 in.
Ⓕ . . . 35 ÷ Ⓔ = 5 in.
Ⓖ . . . Ⓑ − Ⓕ = 1 in.
Ⓗ . . . 25 ÷ Ⓕ = 5 in.
Ⓘ . . . 40 ÷ Ⓗ = 8 in.
Ⓙ . . . 48 ÷ Ⓘ = 6 in.
Ⓚ . . . 66 ÷ Ⓙ = 11 in.
Ⓛ . . . Ⓚ − Ⓕ = 6 in.
Ⓜ . . . 60 ÷ Ⓛ = 10 in.
Ⓝ . . . Ⓛ − Ⓖ = 5 in.
Ⓞ . . . 60 ÷ Ⓝ = 12 in.
Ⓟ . . . 36 ÷ Ⓙ = 6 in.
Ⓠ . . . 24 ÷ Ⓟ = 4 in.
Ⓡ . . . Ⓜ − Ⓠ = 6 in.
Ⓢ . . . 30 ÷ Ⓡ = 5 in.
Ⓣ . . . 20 ÷ Ⓢ = 4 in.
Ⓤ . . . Ⓞ − Ⓣ = 8 in.
Ⓥ . . . 32 ÷ Ⓤ = 4 in.
Length ⓘ is Ⓝ + Ⓥ = 9 in.

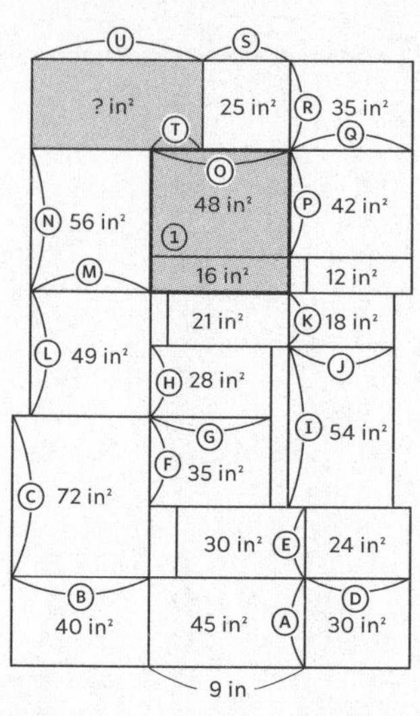

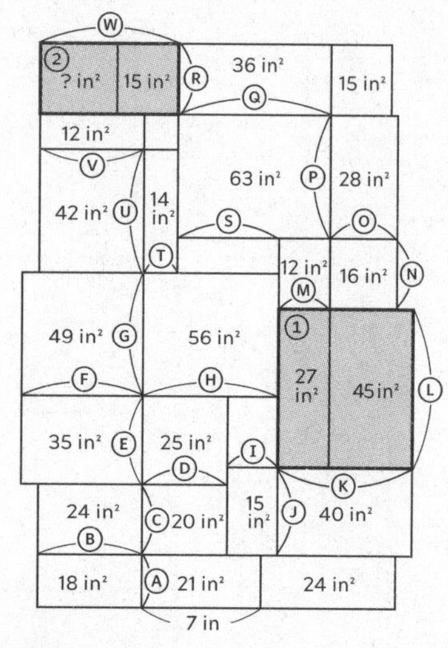

Puzzle 58

Ⓐ . . . 45 ÷ 9 = 5 in.
Ⓑ . . . 40 ÷ Ⓐ = 8 in.
Ⓒ . . . 72 ÷ Ⓑ = 9 in.
Ⓓ . . . 30 ÷ Ⓐ = 6 in.
Ⓔ . . . 24 ÷ Ⓓ = 4 in.
Ⓕ . . . Ⓒ – Ⓔ = 5 in.
Ⓖ . . . 35 ÷ Ⓕ = 7 in.
Ⓗ . . . 28 ÷ Ⓖ = 4 in.
Ⓘ . . . Ⓕ + Ⓗ = 9 in.
Ⓙ . . . 54 ÷ Ⓘ = 6 in.
Ⓚ . . . 18 ÷ Ⓙ = 3 in.
Ⓛ . . . Ⓗ + Ⓚ = 7 in.
Ⓜ . . . 49 ÷ Ⓛ = 7 in.
Ⓝ . . . 56 ÷ Ⓜ = 8 in.
Area ① . . .
48 = 16 = 64 in.^{2}
Ⓞ . . . 64 ÷ Ⓝ = 8 in.
Ⓟ . . . 48 ÷ Ⓞ = 6 in.
Ⓠ . . . 42 ÷ Ⓟ = 7 in.
Ⓡ . . . 35 ÷ Ⓠ = 5 in.
Ⓢ . . . 25 ÷ Ⓡ = 5 in.
Ⓣ . . . Ⓞ – Ⓢ = 3 in.
Ⓤ . . . Ⓜ + Ⓣ = 10 in.
Area ? is Ⓡ × Ⓤ = 50 in.^{2}

Puzzle 59

Ⓐ . . . 21 ÷ 7 = 3 in.
Ⓑ . . . 18 ÷ Ⓐ = 6 in.
Ⓒ . . . 24 ÷ Ⓑ = 4 in.
Ⓓ . . . 20 ÷ Ⓒ = 5 in.
Ⓔ . . . 25 ÷ Ⓓ = 5 in.
Ⓕ . . . 35 ÷ Ⓔ = 7 in.
Ⓖ . . . 49 ÷ Ⓕ = 7 in.
Ⓗ . . . 56 ÷ Ⓖ = 8 in.
Ⓘ . . . Ⓗ – Ⓓ = 3 in.
Ⓙ . . . 15 ÷ Ⓘ = 5 in.
Ⓚ . . . 40 ÷ Ⓙ = 8 in.
Area ① . . .
27 + 45 = 72 in.^{2}
Ⓛ . . . 72 ÷ Ⓚ = 9 in.
Ⓜ . . . 27 ÷ Ⓛ = 3 in.
Ⓝ . . . 12 ÷ Ⓜ = 4 in.
Ⓞ . . . 16 ÷ Ⓝ = 4 in.
Ⓟ . . . 28 ÷ Ⓞ = 7 in.
Ⓠ . . . 63 ÷ Ⓟ = 9 in.
Ⓡ . . . 36 ÷ Ⓠ = 4 in.
Ⓢ . . . Ⓠ – Ⓜ = 6 in.
Ⓣ . . . Ⓗ – Ⓢ = 2 in.
Ⓤ . . . 14 ÷ Ⓣ = 7 in.
Ⓥ . . . 42 ÷ Ⓤ = 6 in.
Ⓦ . . . Ⓣ + Ⓥ = 8 in.
Area ② . . .
Ⓡ × Ⓦ = 32 in.^{2}
Area ? is 32 – 15 = 17 in.^{2}

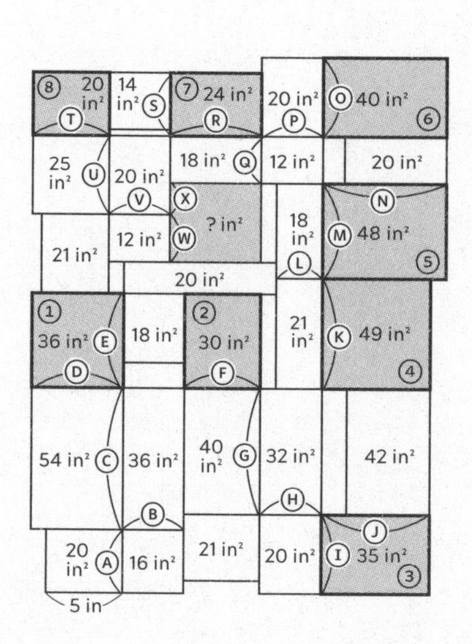

Puzzle 60

Ⓐ . . . 20 ÷ 5 = 4 in.
Ⓑ . . . 16 ÷ Ⓐ = 4 in.
Ⓒ . . . 36 ÷ Ⓑ = 9 in.
Ⓓ . . . 54 ÷ Ⓒ = 6 in.
Ⓔ . . . 36 ÷ Ⓓ = 6 in.
Areas ① and ② are the same height.
Ⓕ . . . 30 ÷ Ⓔ = 5 in.
Ⓖ . . . 40 ÷ Ⓕ = 8 in.
Ⓗ . . . 32 ÷ Ⓖ = 4 in.
Ⓘ . . . 20 ÷ Ⓗ = 5 in.
Ⓙ . . . 35 ÷ Ⓘ = 7 in.
Areas ③ and ④ are the same width.
Ⓚ . . . 49 ÷ Ⓙ = 7 in.
Ⓛ . . . 21 ÷ Ⓚ = 3 in.
Ⓜ . . . 18 ÷ Ⓛ = 6 in.
Ⓝ . . . 48 ÷ Ⓜ = 8 in.
Areas ⑤ and ⑥ are the same width.
Ⓞ . . . 40 ÷ Ⓝ = 5 in.
Ⓟ . . . 20 ÷ Ⓞ = 4 in.
Ⓠ . . . 12 ÷ Ⓟ = 3 in.
Ⓡ . . . 18 ÷ Ⓠ = 6 in.
Ⓢ . . . 24 ÷ Ⓡ = 4 in.
Areas ⑦ and ⑧ are the same height.
Ⓣ . . . 20 ÷ Ⓢ = 5 in.
Ⓤ . . . 25 ÷ Ⓣ = 5 in.
Ⓥ . . . 20 ÷ Ⓤ = 4 in.
Ⓦ . . . 12 ÷ Ⓥ = 3 in.
Ⓧ . . . Ⓤ – Ⓠ = 2 in.
Area ? is Ⓡ × (Ⓧ + Ⓦ) = 30 in.^{2}

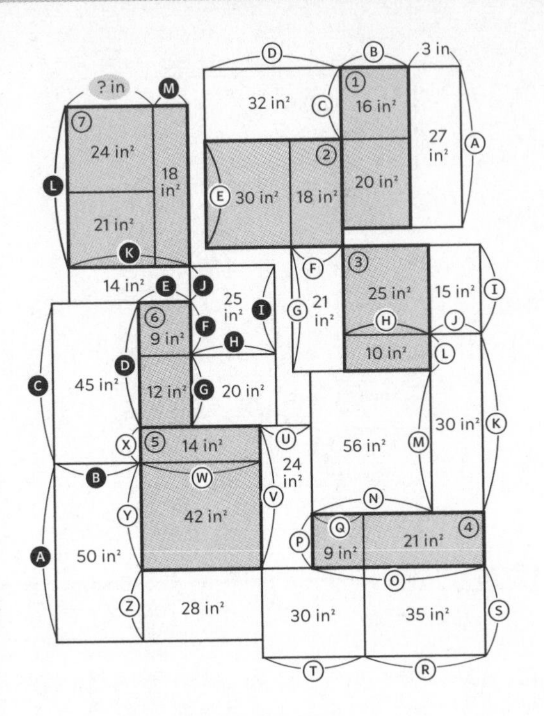

Puzzle 61

Ⓐ . . . 27 ÷ 3 = 9 in.
Area ① . . . 16 + 20 = 36 in.²
Ⓑ . . . 36 ÷ Ⓐ = 4 in.
Ⓒ . . . 16 ÷ Ⓑ = 4 in.
Ⓓ . . . 32 ÷ Ⓒ = 8 in.
Area ② . . . 30 + 18 = 48 in.²
Ⓔ . . . 48 ÷ Ⓓ = 6 in.
Ⓕ . . . 18 ÷ Ⓔ = 3 in.
Ⓖ . . . 21 ÷ Ⓕ = 7 in.
Area ③ . . . 25 + 10 = 35 in.²
Ⓗ . . . 35 ÷ Ⓖ = 5 in.
Ⓘ . . . 25 ÷ Ⓗ = 5 in.
Ⓙ . . . 15 ÷ Ⓘ = 3 in.
Ⓚ . . . 30 ÷ Ⓙ = 10 in.
Ⓛ . . . Ⓖ – Ⓘ = 2 in.
Ⓜ . . . Ⓚ – Ⓛ = 8 in.
Ⓝ . . . 56 ÷ Ⓜ = 7 in.
Ⓞ . . . Ⓙ + Ⓝ = 10 in.
Area ④ . . . 9 + 21 = 30 in.²
Ⓟ . . . 30 ÷ Ⓞ = 3 in.
Ⓠ . . . 9 ÷ Ⓟ = 3 in.
Ⓡ . . . Ⓞ – Ⓠ = 7 in.
Ⓢ . . . 35 ÷ Ⓡ = 5 in.
Ⓣ . . . 30 ÷ Ⓢ = 6 in.
Ⓤ . . . Ⓣ – Ⓠ = 3 in.
Ⓥ . . . 24 ÷ Ⓤ = 8 in.
Area ⑤ . . . 14 + 42 = 56 in.²
Ⓦ . . . 56 ÷ Ⓥ = 7 in.
Ⓧ . . . 14 ÷ Ⓦ = 2 in.
Ⓨ . . . 42 ÷ Ⓦ = 6 in.
Ⓩ . . . 28 ÷ Ⓦ = 4 in.
🅐 . . . Ⓨ + Ⓩ = 10 in.
🅑 . . . 50 ÷ 🅐 = 5 in.
🅒 . . . 45 ÷ 🅑 = 9 in.
🅓 . . . 🅒 – Ⓧ = 7 in.
Area ⑥ . . . 9 + 12 = 21 in.²
🅔 . . . 21 ÷ 🅓 = 3 in.
🅕 . . . 9 ÷ 🅔 = 3 in.
🅖 . . . 12 ÷ 🅔 = 4 in.
🅗 . . . 20 ÷ 🅖 = 5 in.
🅘 . . . 25 ÷ 🅗 = 5 in.
🅙 . . . 🅘 – 🅕 = 2 in.
🅚 . . . 14 ÷ 🅙 = 7 in.
Area ⑦ . . . 24 + 21 + 18 = 63 in.²
🅛 . . . 63 ÷ 🅚 = 9 in.
🅜 . . . 18 ÷ 🅛 = 2 in.
Length ? is 🅚 – 🅜 = 5 in.

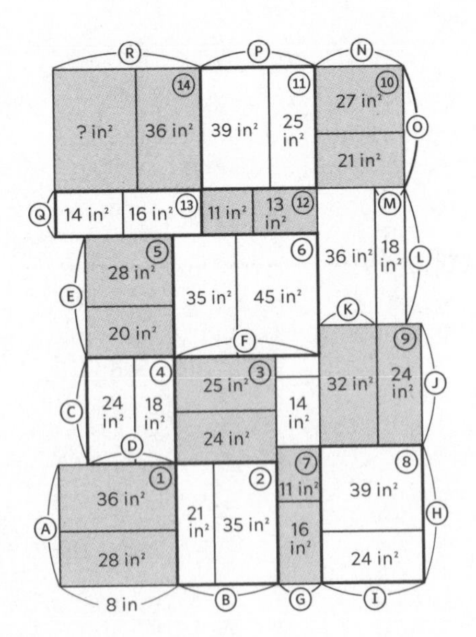

Puzzle 62

Area ① . . .
36 + 28 = 64 in.²
Ⓐ . . . 64 ÷ 8 = 8 in.
Area ② . . .
21 + 35 = 56 in.²
Ⓑ . . . 56 ÷ Ⓐ = 7 in.
Area ③ . . .
25 + 24 = 49 in.²
Ⓒ . . . 49 ÷ Ⓑ = 7 in.
Area ④ . . .
24 + 18 = 42 in.²
Ⓓ . . . 42 ÷ Ⓒ = 6 in.
Area ⑤ . . .
28 + 20 = 48 in.²
Ⓔ . . . 48 ÷ Ⓓ = 8 in.
Area ⑥ . . .
35 + 45 = 80 in.²
Ⓕ . . . 80 ÷ Ⓔ = 10 in.
Ⓖ . . . Ⓕ – Ⓑ = 3 in.
Area ⑦ . . .
11 + 16 = 27 in.²
Ⓗ . . . 27 ÷ Ⓖ = 9 in.
Area ⑧ . . .
39 + 24 = 63 in.²
Ⓘ . . . 63 ÷ Ⓗ = 7 in.
Area ⑨ . . .
32 + 24 = 56 in.²
Ⓙ . . . 56 ÷ Ⓘ = 8 in.
Ⓚ . . . 32 ÷ Ⓙ = 4 in.
Ⓛ . . . 36 ÷ Ⓚ = 9 in.
Ⓜ . . . 18 ÷ Ⓛ = 2 in.
Area ⑩ . . .
27 + 21 = 48 in.²
Ⓝ . . . Ⓚ + Ⓜ = 6 in.
Ⓞ . . . 48 ÷ Ⓝ = 8 in.
Area ⑪ . . .
39 + 25 = 64 in.²
Ⓟ . . . 64 ÷ Ⓞ = 8 in.
Area ⑫ . . . 11 + 13 = 24 in.²
Ⓠ . . . 24 ÷ Ⓟ = 3 in.
Area ⑬ . . . 14 + 16 = 30 in.²
Ⓡ . . . 30 ÷ Ⓠ = 10 in.²
Area ⑭ . . . 10 × Ⓞ = 80 in.²
Area ? is 80 – 36 = 44 in.²

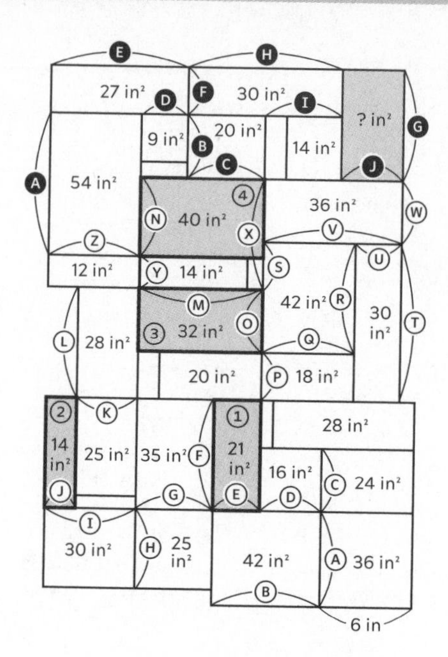

Puzzle 63

Ⓐ . . . 36 ÷ 6 = 6 in.
Ⓑ . . . 42 ÷ Ⓐ = 7 in.
Ⓒ . . . 24 ÷ 6 = 4 in.
Ⓓ . . . 16 ÷ Ⓒ = 4 in.
Ⓔ . . . Ⓑ − Ⓓ = 3 in.
Ⓕ . . . 21 ÷ Ⓔ = 7 in.
Ⓖ . . . 35 ÷ Ⓕ = 5 in.
Ⓗ . . . 25 ÷ Ⓖ = 5 in.
Ⓘ . . . 30 ÷ Ⓗ = 6 in.
Areas ① and ② are the same height.
Ⓙ . . . 14 ÷ Ⓕ = 2 in.
Ⓚ . . . Ⓘ − Ⓙ = 4 in.
Ⓛ . . . 28 ÷ Ⓚ = 7 in.
Ⓜ . . . Ⓔ + Ⓖ = 8 in.
Areas ③ and ④ are the same width.
Ⓝ . . . 40 ÷ Ⓜ = 5 in.
Ⓞ . . . 32 ÷ Ⓜ = 4 in.
Ⓟ . . . Ⓛ − Ⓞ = 3 in.
Ⓠ . . . 18 ÷ Ⓟ = 6 in.
Ⓡ . . . 42 ÷ Ⓠ = 7 in.
Ⓢ . . . Ⓡ − Ⓞ = 3 in.
Ⓣ . . . Ⓟ + Ⓡ = 10 in.
Ⓤ . . . 30 ÷ Ⓣ = 3 in.
Ⓥ . . . Ⓠ + Ⓤ = 9 in.
Ⓦ . . . 36 ÷ Ⓥ = 4 in.
Ⓧ . . . Ⓢ + Ⓦ = 7 in.
Ⓨ . . . Ⓧ − Ⓝ = 2 in.
Ⓩ . . . 12 ÷ Ⓨ = 6 in.
🅐 . . . 54 ÷ Ⓩ = 9 in.
🅑 . . . 🅐 − Ⓝ = 4 in.
🅒 . . . 20 ÷ 🅑 = 5 in.
🅓 . . . Ⓜ − 🅒 = 3 in.
🅔 . . . Ⓩ + 🅓 = 9 in.
🅕 . . . 27 ÷ 🅔 = 3 in.
🅖 . . . 🅑 + 🅕 = 7 in.
🅗 . . . 30 ÷ 🅕 = 10 in.
🅘 . . . 🅗 − 🅒 = 5 in.
🅙 . . . Ⓥ − 🅘 = 4 in.
Area ? is 🅖 × 🅙
= 28 in.²

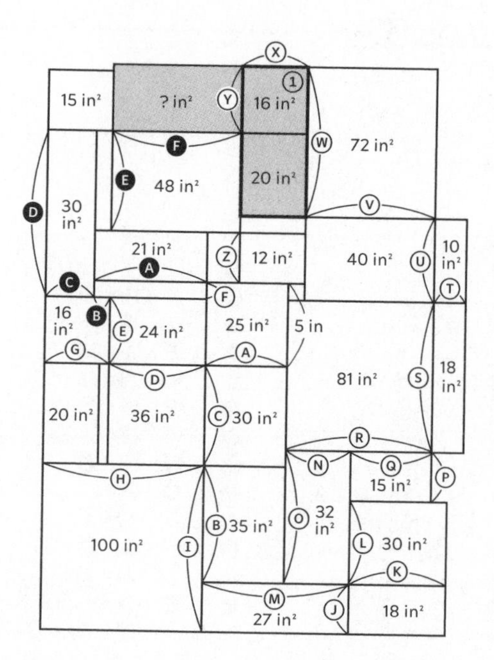

Puzzle 64

Ⓐ . . . 25 ÷ 5 = 5 in.
Ⓑ . . . 35 ÷ Ⓐ = 7 in.
Ⓒ . . . 30 ÷ Ⓐ = 6 in.
Ⓓ . . . 36 ÷ Ⓒ = 6 in.
Ⓔ . . . 24 ÷ Ⓓ = 4 in.
Ⓕ . . . 5 − Ⓔ = 1 in.
Ⓖ . . . 16 ÷ Ⓔ = 4 in.
Ⓗ . . . Ⓓ + Ⓖ = 10 in.
Ⓘ . . . 100 ÷ Ⓗ = 10 in.
Ⓙ . . . Ⓘ − Ⓑ = 3 in.
Ⓚ . . . 18 ÷ Ⓙ = 6 in.
Ⓛ . . . 30 ÷ Ⓚ = 5 in.
Ⓜ . . . 27 ÷ Ⓙ = 9 in.
Ⓝ . . . Ⓜ − Ⓐ = 4 in.
Ⓞ . . . 32 ÷ Ⓝ = 8 in.
Ⓟ . . . Ⓞ − Ⓛ = 3 in.
Ⓠ . . . 15 ÷ Ⓟ = 5 in.
Ⓡ . . . Ⓝ + Ⓠ = 9 in.
Ⓢ . . . 81 ÷ Ⓡ = 9 in.
Ⓣ . . . 18 ÷ Ⓢ = 2 in.
Ⓤ . . . 10 ÷ Ⓣ = 5 in.
Ⓥ . . . 40 ÷ Ⓤ = 8 in.
Ⓦ . . . 72 ÷ Ⓥ = 9 in.
Area ① . . .
16 + 20 = 36 in.²
Ⓧ . . . 36 ÷ Ⓦ = 4 in.
Ⓨ . . . 16 ÷ Ⓧ = 4 in.
Ⓩ . . . 12 ÷ Ⓧ = 3 in.
🅐 . . . 21 ÷ Ⓩ = 7 in.
🅑 . . . 🅐 − Ⓓ = 1 in.
🅒 . . . Ⓖ − 🅑 = 3 in.
🅓 . . . 30 ÷ 🅒 = 10 in.
🅔 . . .
🅓 − Ⓕ − Ⓩ = 6 in.
🅕 . . . 48 ÷ 🅔 = 8 in.
Area ? is Ⓨ × 🅕
= 32 in.²

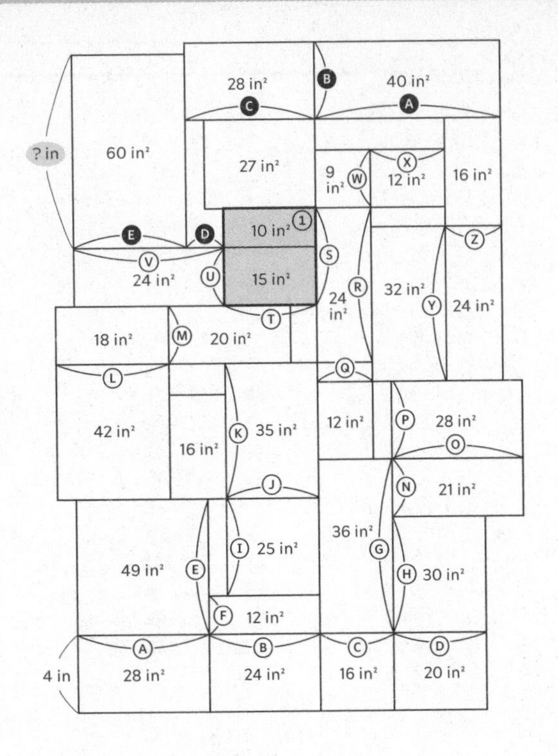

Puzzle 65

Ⓐ . . . 28 ÷ 4 = 7 in.
Ⓑ . . . 24 ÷ 4 = 6 in.
Ⓒ . . . 16 ÷ 4 = 4 in.
Ⓓ . . . 20 ÷ 4 = 5 in.
Ⓔ . . . 49 ÷ Ⓐ = 7 in.
Ⓕ . . . 12 ÷ Ⓑ = 2 in.
Ⓖ . . . 36 ÷ Ⓒ = 9 in.
Ⓗ . . . 20 ÷ Ⓓ = 6 in.
Ⓘ . . . Ⓔ − Ⓕ = 5 in.
Ⓙ . . . 25 ÷ Ⓘ = 5 in.
Ⓚ . . . 35 ÷ Ⓙ = 7 in.
Ⓛ . . . 42 ÷ Ⓚ = 6 in.
Ⓜ . . . 18 ÷ Ⓛ = 3 in.
Ⓝ . . . Ⓖ − Ⓗ = 3 in.
Ⓞ . . . 21 ÷ Ⓝ = 7 in.
Ⓟ . . . 28 ÷ Ⓞ = 4 in.
Ⓠ . . . 12 ÷ Ⓟ = 3 in.
Ⓡ . . . 24 ÷ Ⓠ = 8 in.
Ⓢ . . . Ⓡ − Ⓜ = 5 in.
Area ① . . .
$10 + 15 = 25$ in.²
Ⓣ . . . 25 ÷ Ⓢ = 5 in.
Ⓤ . . . 15 ÷ Ⓣ = 3 in.
Ⓥ . . . 24 ÷ Ⓤ = 8 in.
Ⓦ . . . 9 ÷ Ⓠ = 3 in.
Ⓧ . . . 12 ÷ Ⓦ = 4 in.
Ⓨ . . . 32 ÷ Ⓧ = 8 in.
Ⓩ . . . 24 ÷ Ⓨ = 3 in.
🅐 . . . Ⓠ + Ⓧ + Ⓩ
= 10 in.
🅑 . . . 40 ÷ 🅐 = 4 in.
🅒 . . . 28 ÷ 🅑 = 7 in.
🅓 . . . 🅒 − Ⓣ = 2 in.
🅔 . . . Ⓥ − 🅓 = 6 in.
Length (?) is 60 ÷ 🅔
= 10 in.

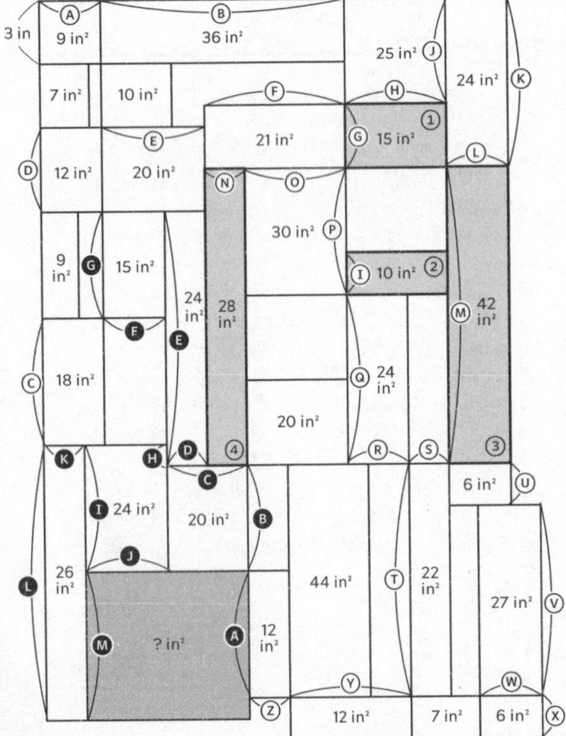

Puzzle 66

Ⓐ . . . 9 ÷ 3 = 3 in.
Ⓑ . . . 36 ÷ Ⓐ = 12 in.
Ⓒ . . . 18 ÷ Ⓐ = 6 in.
Ⓓ . . . 12 ÷ Ⓐ = 4 in.
Ⓔ . . . 20 ÷ Ⓓ = 5 in.
Ⓕ . . . Ⓑ − Ⓔ = 7 in.
Ⓖ . . . 21 ÷ Ⓕ = 3 in.
Ⓗ . . . 15 ÷ Ⓖ = 5 in.
Areas ① and ② are the same width.
Ⓘ . . . 10 ÷ Ⓗ = 2 in.
Ⓙ . . . 25 ÷ Ⓗ = 5 in.
Ⓚ . . . Ⓖ + Ⓙ = 8 in.
Ⓛ . . . 24 ÷ Ⓚ = 3 in.
Ⓜ . . . 42 ÷ Ⓛ = 14 in.
Areas ③ and ④ are the same height.
Ⓝ . . . 28 ÷ Ⓜ = 2 in.
Ⓞ . . . Ⓕ − Ⓝ = 5 in.
Ⓟ . . . 30 ÷ Ⓞ = 6 in.
Ⓠ . . . Ⓜ − Ⓟ = 8 in.
Ⓡ . . . 24 ÷ Ⓠ = 3 in.
Ⓢ . . . Ⓗ − Ⓡ = 2 in.
Ⓣ . . . 22 ÷ Ⓢ = 11 in.
Ⓤ . . . 6 ÷ Ⓛ = 2 in.
Ⓥ . . . Ⓣ − Ⓤ = 9 in.
Ⓦ . . . 27 ÷ Ⓥ = 3 in.
Ⓧ . . . 6 ÷ Ⓦ = 2 in.
Ⓨ . . . 12 ÷ Ⓧ = 6 in.
Ⓩ . . . Ⓞ + Ⓡ − Ⓨ = 2 in.
🅐 . . . 12 ÷ Ⓩ = 6 in.
🅑 . . . Ⓣ − 🅐 = 5 in.
🅒 . . . 20 ÷ 🅑 = 4 in.
🅓 . . . 🅒 − Ⓝ = 2 in.
🅔 . . . 24 ÷ 🅓 = 12 in.
🅕 . . . Ⓔ − 🅓 = 3 in.
🅖 . . . 15 ÷ 🅕 = 5 in.
🅗 . . . 🅔 − Ⓒ − 🅖 = 1 in.
🅘 . . . 🅑 + 🅗 = 6 in.
🅙 . . . 24 ÷ 🅘 = 4 in.
🅚 . . . Ⓐ + 🅕 − 🅙 = 2 in.
🅛 . . . 26 ÷ 🅚 = 13 in.
🅜 . . . 🅛 − 🅘 = 7 in.
Area (?) is 🅜 × (🅒 + 🅙)
= 56 in.²